THE
COCKTAIL
GUY

RICH WOODS

THE COCKTAIL GUY

Infusions, distillations
and innovative combinations

PAVILION

CONTENTS

MY KIND OF DRINKS

The assortment of memories that make up my internal narrative of childhood is patched together from strong recollections of the joy and laughter of good times, the warmth and closeness of family, the feeling of embarking on great adventures and, overarching it all, delicious, fresh and simply but beautifully prepared food. Whether I was spending the summer at my grandparents' home, shucking peas or picking fruit from my granddad's allotment; or vacationing near the sea, where my sister and I would join the hordes of other children searching nearby hay barns for chicken eggs (the honeyed aroma of hay still reminds me of those days), my earliest and strongest memories centre around food and, even today, the faintest whiff of a particular smell or the hint of a familiar flavour is enough to transport me straight back to my youth.

I was incredibly fortunate when growing up that the food that I ate wasn't only nutritious, but also packed with freshness and flavour to such a degree that early on, and unbeknownst to me at the time, I was laying the foundations for my own personal flavour bank – a tool that influences and informs my creative process and forms the basis of every new drink I make. Family dinners were often eaten al fresco in the garden of my childhood home and the flavours of the delicious food my mother had prepared combined with the aromatic smells that wafted over our country garden – fragrant fresh herbs, plump tomatoes growing on the vine, the heady perfume of the flower garden – came together to provide a sensory feast and every fragrance, texture and flavour was logged by my subconscious ready to be recalled later.

Back when I started frequenting bars, cocktail menus were often bible-thick tomes that listed volumes of drinks, rather than concentrating on a compact and concise offering. It was common to see whole pages offering various fruit-flavoured Martinis or collinses listed amongst their offering – each one just slightly tweaked to expand the list even further. Guests weren't looking for boundary pushing or truly memorable drinks and many barmen were just seeking to emulate what had come before: looking backwards, not forwards.

Luckily things have changed and drinking in the modern day is about quality, experience and information. The modern drinker is a more experience-led individual, wanting encounters that offer a shift of polarity in tastes and trends. They are interested in drinks that move perception around challenging ingredients in moments that stimulate provocation. It is these 'a-ha' moments that afford me the greatest joy when presenting a new drink; watching a guest's face change from expectation, to surprise, to excitement, all in the matter of a few seconds as they taste their first sip. Drinking, after all, is about escapism; the intangible over the transactional.

In this book, I hope that you find drinks that will both engage your senses and ignite your creative desire. Included are those that are greatly simple and others that are simply great (but might require a bit more effort on your part), but with all the drinks, the approach is always the same. My aim is to take you step-by-step through the creative process and also to enlighten you as to where the ideas behind these drinks came from. After all, they say the devil is in the detail.

Whether you follow my recipes verbatim or use them to inspire your own creations, I hope that you find something new and exciting within these pages and enjoy experiencing my kind of drinks. Just remember, no recipe is a proven perfect drink. Experiment with various measurements and ratios as your palate is personal to you, so adapt the drinks and recipes as needed. There is no such thing as failure – each misstep is another move forward on the path to success. Giving up, however, is a conscious decision, so always question and always take notes on your findings. Write down all your ideas and note even the unsuccessful ones. You never know when an occasion or smell or moment will spark a new creation.

Above all, create something different.

RICH

UNDERSTANDING FLAVOUR

Before we go any further in the book, I'd like to show you by way of a test the importance of smell and its relevance to flavour. This is a great little exercise that also highlights why, when presenting a drink, a good garnish is there to do far more than look pretty.

I'm sure that everyone reading this has suffered a cold at some point in their life – after all, it's not called the common cold for nothing. A bunged-up nose is horrible: endless blowing, only to be stuffed up again seconds later. And then there's the distinct loss of your sense of taste. Bread and soup are rendered tasteless and under-seasoned. Tea tastes weak, biscuits dry. Annoying as it is, this loss of taste is perfectly normal and easily explained.

So, to the test: try pinching your nose closed and then put some food into your mouth. The usual burst of flavour will be absent. This is why the nose is so important and also why, when we have a cold and a blocked nose, we find it difficult to appreciate food.

Taste is an experience restricted to five elements. Detected as salt, bitter, sour, sweet and umami – all of which are descriptors used when defining a drink. Though there are thousands of receptors on our tongue designed to identify and respond to these elements, these are simple tastes that are hard-wired from birth, whereas the flavours that we learn through smell are subjective, and therefore a more powerful property. Smells are subject to individual preferences and difference. In short, without the addition of smell, taste is limited.

The olfactory bulb organizes smell molecules in the same way the retina organizes and displays visual fields, then sends signals through the optic nerve to the brain. Smell is the only sense that connects directly to the brain's limbic system – which recalls memories, emotion and perception. This direct pathway gives scent its power and it is why we form such strong bonds to things that taste and smell good.

I love food and love eating. Not only because every mouthful is packed with flavours, but also because of the circumstances in which we eat – where, when and with whom. Perhaps because of this, I approach cocktail-making the same way that a chef does food. My focus is on flavour.

The process is quite simple. Starting with a dominant flavour, a blank page and a pen, I then start pairing other ingredients with the primary flavour in a process referred to as flavour bouncing. This may sound basic but trust me, if it works for food, it works for drinks – most of the time! I then look for ingredients that connect two or more elements to each other – these are called bridging flavours. It is only then that I think about what most people consider the main act – the spirit or liquor. This to me is a bridging ingredient – it works to tie multiple elements together and allows me to choose the right spirit for a particular drink.

Creating a drink is more than simply throwing liquid into a glass. I look at each drink as an experience. From the moment a menu is written or a drink is described, we are influencing how the brain reacts and how the drink is perceived. The way a cocktail is presented, from the vessel it is served in to its weight, colour and temperature, are all considerations – after all, you never get a second chance to make a first impression.

The mouthfeel and texture of a drink are also vitally important and I am continuously looking into how I can better an experience and make a drink more enjoyable. In this way, I regularly mislead, misguide, manipulate and flirt with a drink's boundaries, but I always ensure that the cornerstones of flavour are respected. The trick is not to stray too far from the foundations of a cocktail, to create something surprising that still tastes delicious.

Nobody drinks a cocktail to satisfy their thirst. We drink to be removed from our daily lives and surrender ourself to an experience. A good cocktail is both a luxury and an escape and should be worthy of both of these labels.

A NOTE ON THE RECIPES

The recipes in this book range in difficulty on a scale of
1–5 which is clearly marked on each recipe. Many of the
recipes require you to infuse elements with flavour, which
often requires minimal effort, but a little patience. Again,
the preparation time (including the time to prepare any
infusions) is clearly marked on each recipe.

The suggested times are based on maximum infusion
times. If you prefer a drink with a less pronounced flavour
you can reduce the infusion time. Best practice would be to
taste as you go along.

The recipes in the bulk of the book are perfectly
achievable for the keen home bartender, but most do
require a little forward planning. There are various ways
of infusing a spirit with flavour (these are described
in detail on pages 22–23) and if you do not have the
equipment required for my recommended method (i.e.
vacuum packing and cooking by sous vide), you can
try one of the other methods for a similar (though less
pronounced) result.

The recipes in the Iconoclastic chapter (see page 124)
are by far the most difficult and will be beyond the reach
of the home bartender who doesn't have access to the
very expensive equipment needed to create their own
distillations at home and the licences required to distil
their own alcohol. This chapter is brief, but includes some
of my most iconic recipes, so I wanted to include it to
give a full picture of my creative process.

Where possible I have not stipulated a specific brand of
alcohol to use within a drink and any neutral-flavoured
spirit will work perfectly well, but in some instances,
where I do feel that the aromatics used in the creation of
a specific spirit tie in especially well with a cocktail and will
elevate the final result, I stipulate a specific brand in the
ingredients list. It's up to you if you want to go to the
extra effort (and expense) of procuring this particular
variety of spirit, but the results won't be quite the same if
you do not.

*Rules and regulations regarding distilling vary from country to
country even if you are not planning on selling your distillations.
Please check out local laws before embarking on distilling your
own drinks.*

FENNEL FIZZ

DISTILLED

SOUS VIDE

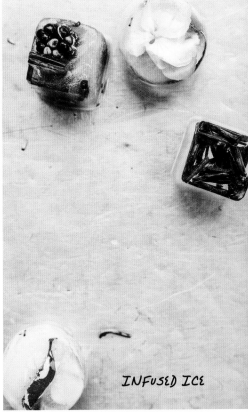

INFUSED ICE

STRAINER

MICRO SCALES

THE BASICS

ESSENTIAL TOOLS

Whilst it is true that a good tradesman never blames his tools, he who does not invest in the future of his trade is a bad one. In this day and age, when the high street is flooded with good-quality and affordable products, there is simply no excuse for not having quality equipment – but quality does not necessarily have to mean expensive. There are many options and so many places to look. Building up your equipment should never be a budget-blowing exercise, and is something that you can add to week by week or month by month.

If you search the web you can find a great variety of equipment, from the most basic of shakers to novelty straws. You can also find retro pieces that can date back several decades – these tend to be a bit pricier but can provide a great talking point. Personally, I've always found Cocktail Kingdom (see Suppliers, page 170) to be a great place. They stock hundreds of pieces of equipment and ship to almost anywhere in the world. Check page 170 for links to other recommended suppliers.

When I am asked to do demos or videos that focus on entertaining at home, I often bring equipment bought from the high street – highlighting how easy and affordable quality can be. You don't need to be a bar professional to create beautiful cocktails and many mainstream department stores now have collections of home entertaining paraphernalia, including shakers, measures or jiggers, stirrers, bar spoons and ice moulds. But with so much available, just what are the essentials?

NOTEBOOK AND PEN

Of all the kit in my armoury, a notebook and pen are the only things that travel everywhere with me. From ideas for flavour combinations, new infusions or techniques, inspiration is everywhere and it is vital that I make note of any spark of an idea as it comes, otherwise it is quickly in danger of being lost forever. Inspiration can strike at the most inopportune of times, whether I'm struck by a distinctive aroma on the journey to work, have an epiphany walking down the street or try an unusual flavour combination whilst out for dinner, and every time: out comes the book. Note down all your ideas and trials – even the bad ones (that way you'll remember not to repeat them).

ICE MOULDS

To put it bluntly, crappy ice makes for crappy drinks. Whether you're a keen home bartender or a professional cocktail slinger, good ice moulds are an essential part of your kit.

Moulds come in various shapes and sizes, from the standard cube to a perfect sphere. They can be purchased almost anywhere – from Amazon to kitchen supply shops – and a gorgeous piece or pieces of ice, makes a drink pop! (For more on ice, see page 30.)

SHAKERS

There are several common styles of shaker available and now, thanks to the many lifestyle magazines, they are available in most finishes of metal – from the classic stainless to copper and gold plated, and everything in-between. The classic three-piece or cobbler shaker is a great starting point to a collection and comes in a range of sizes. My personal preference is tin-on-tin. A set or two of sturdy two-piece stainless shakers are great, durable and reliable as well as also doubling as a stirring vessel. Never use your shakers to strain a drink as if you are cracking an egg and don't use one shaker placed inside the other, for the same result.

STRAINERS

A good strainer makes all the difference to producing a sharp-looking drink. While it sounds trivial, it is vital not to underestimate the importance of an exceptional double strain but it is a fine balance and your equipment plays a key role. A good strainer should be fine enough to remove any ice shards left over from shaking or mixing a drink, but not so fine as to remove air bubbles and thus texture. As with all equipment – you get what you pay for.

Single or Hawthorne Strainers – The flat-ended, coiled one
Essential to single strain a stirred drink and damn tricky to double strain without one (I've actually seen someone do this). Remove the tightly coiled spring and use as a whisk in a dry-shaken egg-white drink for greater texture.

Double or Fine Sieve Strainer – The mesh one
Stay away from the small cheap variety, which often have mesh so fine that drinks suffer from its use. Opt instead for a decent size with a medium mesh which is adequate to hold back unwanted ice shards, fruit scraps, pith or seeds, whilst allowing aeration to pass.

Julep Strainer – The perforated, bowl shaped one.
Originally designed to hold back crushed ice and mint in the popularized Julep cocktail in the days before straws were available. Used frequently for stirred drinks.

I personally opt for the Hawthorne option, minimizing the need for all three types.

BAR SPOONS*

Used for a variety of techniques, from the obvious stirring and churning, to layering, measuring and garnishing. The bar spoon is an integral part of the bartender's weaponry. Some might come with a coin- or disk-shaped flat end or a trident-style fork, for grabbing cocktail olives or onions. It is worth noting that the measurement of these various styles differs from 2.5ml up to 5ml depending on the style or where it is purchased and when making drinks that require a perfect pour or ratio, the difference in a millilitre is as good as a miss.

MEASURES/JIGGERS

Have you ever sipped on a badly made Martini or Manhattan, poured straight from the bottle? Accurate measuring can make or break a cocktail. Consistency creates a standard that everyone is entitled to expect. Often when making cocktails, such as Manhattans or Negronis, I batch in larger quantities. This makes measuring a small amount much easier when scaling up a recipe.

When it comes to measures or jiggers, I opt for the double-ended. Depending on which side of the Atlantic I'm working at the time, this may be either 25 ml/50 ml or 1 fl oz/2 fl oz.

Ensure when you measure, you do so to the mark. If a recipe calls for a specific measurement – give it such.

JUICERS

Be it a hand-held hinged 'elbow' style juicer or the more traditional citrus squeezer, these are great for personal serves but not so good for larger groups. Some of the recipes included here require quantities that will leave you with aching wrists after squeezing all you need. In cases like this, an electric juicer is recommended.

MUDDLERS

Similar in use and appearance to a pestle, the primary role of this useful tool is for the juicing of citrus fruits that require the extraction of the oils located in their skins, as well as cracking spices and squeezing fruits. Opt for a solid number with a silicone end. The wooden varieties can suffer splinter damage as well as being buggers to clean so are best avoided.

*

In the following recipes a bar spoon has been measured as 3 ml. This is the standard flat-end European bar spoon or the equivalent to a standard tea spoon.

01

02

03

04

05

06a

06b

07

08

KNIVES/PEELERS

Have you ever asked a chef how long they have worked with the same knives? Do it. You might be surprised. A great knife can cost as much as pair of designer trainers, but it will last you a hell of a lot longer. It's not essential to spend a fortune, however, and you can take your time to build up your collection, upgrading your knives as your passion and expertise develops (we've all shelled out loads of cash when embarking on a hobby only to have the expensive items gathering dust at the back of a cupboard six months later). As a starting point, these are the knives I would never be without:

Paring Knife – For the bulk of the work, including zesting.

Chef's Knife – For heavy-duty cutting and slicing, in particular for larger fruits and veg.

Channel Knife – To create a garnish of long spiral-like zest in a vesper-style drink. Choose one with a 2-in-1 citrus grater for scraping away at zests.

GLASSWARE

A great collection of glassware is always important and gives your drinks that special touch. Whilst the number of different shapes and sizes of glasses on the market are too numerous to mention, the bulk of recipes in this book use the following styles.

Each recipe comes with a suggested style of glass, but don't feel you have to go out and buy new glassware just to make these drinks. See page 170 for suggested suppliers if you do want to follow my recommendations.

 Coupette or Large Martini (approximately 220–240 ml/7½–8 fl oz)

 Small Martini (approximately 130–150 ml/ 4½–5 fl oz)

 Large Wine Glass

 Champagne Flute

 Tasting or Liqueur Glass

 Collins (approximately 300–325 ml/ 10–11 fl oz) or Highball (approximately 350–400 ml/12–14 fl oz)

 Rocks or Old Fashioned Glass

 Sundae Glass

ADDITIONAL EQUIPMENT

As the approach to cocktailing becomes more creative, it is increasingly necessary to supplement the basic kit with more specialist tools. The equipment listed below is used to create the many infusions and distillations that form a core part of my cocktail recipes and are generally used in the construction of the base ingredients of a cocktail, rather than when actually mixing the drink itself.

VEG JUICER

Great for juicing large quantities of vegetables or fruit, such as celery, cucumber, pumpkin or apples.

INFUSION JARS OR SEALABLE CONTAINERS

Essential for infusions and large batches of cocktails, resting ingredients and storing dry goods.

PIPETTE BOTTLES

Great for decanting your concoctions into, such as bitters, solutions and tinctures etc.

SCALES

For measuring. Obvs. As well as regular digital scales, also consider investing in a set of micro scales; these record measurements from a little as 0.1 g and are small enough to fit in your pocket.

MEASURING FLASKS/CYLINDERS

For larger batches and mixes that require measuring anything greater than the capacity of a standard jigger. I have 250 ml/9 fl oz, 500 ml/18 fl oz and 1 litre/1¾ pints varieties in my kit.

GRATER

From nuts to nutmeg, a small grater is always a great addition to any bar kit.

FUNNEL

For quick, mess-free decanting and for holding muslin or coffee filters in order to strain your infusions.

CREAM WHIPPERS

Great for rapid infusing, by using N_2O (nitrous oxide) to inject micro bubbles into the canister and transfer flavour from a solid ingredient or ingredients into your desired liquid.

LARGE SIEVE/CHINOIS

Like a normal sieve, but bigger! A chinois or large sieve is useful when making batches of drinks or for larger scales or proportions.

BLENDER

A blender is a great piece of kit for pulsing a multitude of ingredients together before filtering them, as well as making great frappé-style daiquiris.

Personally, I opt for a bullet-style blender as their small multiple blades chop though robust ingredients more easily and quickly.

STICK OR IMMERSION BLENDER

Great for blending smaller amounts. More versatile and transportable than a traditional blender.

MUSLIN/COFFEE FILTERS

For general filtering I tend to use coffee filters or a muslin-lined sieve (strainer). When trying to filter smaller particles, I use a very fine 100-micron superbag with a non-absorbent nylon filter. Great for making highly flavoured but almost clear liquids and clarified juices.

COLD DRIP

The cold drip is a three-level piece of kit that is used to gently regulate the speed of water that interacts with fresh coffee before being collected in a receiving vessel. I use this for the Ristretto Negroni (see page 83).

DEHYDRATOR

A bench-top piece of kit that removes moisture through dehydration. Great for ingredients such as citrus wheels or tomato slices. It can also be used to dehydrate liquors with a high sugar content, such as Campari.

VACUUM MACHINE

Used to remove the air from vacuum bags prior to sealing them. This gives a greater liquid to infusion contact area and better interaction. Airtight bags also ensure none of the flavours that you are trying to develop can escape.

WATER BATH OR BAIN MARIE

As you read through the recipes in the following pages, you'll notice I do a lot of water-bath cooking under vacuum (sous vide). This piece of kit allows you to control the temperature of the water surrounding your vacuum bag, to within 0.1°C (0.18°F), preventing any heat damage to either the liquid or flavouring.

ROTARY EVAPORATOR

A device predominantly used in chemical laboratories but in recent years, more and more in both the kitchen and bar. It uses vacuum to control air pressure for the efficient and gentle removal of solvents and distillates by way of evaporation. For more information on vacuum evaporation, see page 154.

TECHNIQUES

THE SHAKE

Everyone's technique varies and the best advice here would be to go with whichever approach you feel most comfortable with. Providing you use enough ice, drive your shake with enough force and give your shake the appropriate amount of time, you will end up with a well-shaken drink of around –7°C to –5°C (19°F to 23°F). After around 10–12 seconds, your drink will reach what is called thermal equilibrium – this is the point at which your drink cannot get any colder (unless possibly both the alcohol and the glass you are serving the drink in had been stored in the freezer).

Shaking a drink aerates it and the minute air bubbles produced by this action add texture to your drink. So, when you have finished, it is important to strain and serve your drink as soon as possible.

THE DRY SHAKE

This is a form of pre-shake shake –shaking without ice, usually with egg or cream, to mix and emulsify varying ingredients prior to re-shaking with ice. You can add the small coiled spring of your strainer to the tin, which will act as a small whisk (remember to remove the coil and wash before re-attaching it to the strainer). Fill your shaker with good-quality cube ice and apply your desired shake.

Alternatively, you could shake all your ingredients as you would normally – with ice. Strain the whole mix into one half of your shaker, discard the ice and then apply the dry shake. This practice is known as the 'reverse dry shake'. Just remember, whichever way you do it, ensure you double strain.

THE STIR

Stirring, like shaking, has two benefits. The first is to chill the cocktail and the second is to dilute it. A well-chilled and diluted drink is best served immediately after it has been stirred for around 30 seconds. That said, I always tend to stir a Martini or Manhattan for a little longer as I find that an extra ten seconds gives a greater release in aromatics – but this is purely personal. In any case, always use enough ice and, if you're using a shaker tin to stir, use a large one – it holds more ice and will leave less space in the tin for the ice to start to melt and dilute the drink any further.

For recipes that call for stirring and consist of ingredients other than alcohol – say, for example, a clarified juice – I tend to stir for less than 30 seconds, as this is enough time to chill a drink adequately without diluting it too far.

STRAIN

To separate solids from liquid at the end of an infusion, or to separate ice or fruit sediment from liquid in a cocktail (see Essential Tools, pages 14–15).

MUDDLE

To juice, crack or even lightly bruise an ingredient to encourage the release of more intensely aromatic qualities and oils.

BUILD

To build ingredients upon one another – normally in a glass. This may well call for these ingredients to be mixed together afterwards.

INFUSE

To infuse a liquid is to impart additional flavour by way of adding ingredients to improve a final product. Most commonly this is in the form of fruits or spices infused into a spirit for a period of time. Infusion times vary and depend on both the type of ingredient used and its ability to release flavours. There are various types of infusing and all have varying results.

Standard Infusion

The cheapest and most convenient way of infusing where an ingredient is simply added to a non-reactive container, covered with alcohol and left to rest. It is then filtered once the correct flavour has been achieved.

This is far and away the simplest of the infusion techniques and little thought is required between combining the ingredients and filtering. Though it is possible to get great results from this method, some ingredients respond better to more involved techniques.

Vacuum Infusion

Essentially the same as the above with the addition of a vacuum, i.e. sealing the infusion in an airtight bag. The benefit of this is that there is no chance of any delicate flavours escaping or erroneous ones finding their way in. It's a sterile process and sees more interaction between the liquid and the infusion, leading to better results.

Sous Vide Infusion

This is an additional step to vacuum infusion (see above). To cook sous vide is to heat vacuum-sealed food products in a very carefully temperature-controlled water bath (for more information on sous-vide machines, see below). Essentially you are cooking the contents of your vacuum-sealed bag in a bath of water, controlled to within 0.1°C (0.18°F) and to a recorded and set time. It's faster than a vacuum infusion and I find that I achieve the best flavours from this technique.

Rapid Infusion

As the name suggests, this is a quick infusion that is produced using nitrous oxide (N_2O) and a soda siphon bottle. The siphon is partly filled with a liquid – normally alcohol, but also water or oil – and your chosen aromatic solids. Flavourings are rapidly transferred into the liquid by injecting pressure into the siphon. The pressure forces an infusion with the solids. Upon releasing the gas, tiny bubbles escape the solids and flavour with them, which in turn infuses into the liquid.

CLARIFY

More and more bartenders are using the technique of clarification. And it is one of the simplest. It requires no specific equipment, so can be easily achieved at home with just a little time and effort.

The idea is to clarify a liquid by separating it from its solids, thus creating a near translucent, flavourful liquid. To do this, a gelling agent is added to the liquid, allowing any small fibrous materials contained within it to cling together. The mix is then chilled and frozen. Once frozen, the mix is then melted and filtered through a fine cloth to clarify the liquid.

Personally, I find when clarifying a juice the best results come from only cooking part of the liquid – rather than the total amount. (Gels such as agar agar need to be heated to reach their potential.) Divide the mixture into two and add the gelling agent to the 'cooked' part, ensuring it is thoroughly mixed in, then remove the warm mix from the heat source and add the remaining 'chilled' half. Thoroughly mix and allow to chill to room temperature before freezing.

FILTER

To pass a liquid – normally an infusion – through a tightly woven fabric for the purpose of removing solids.

DEHYDRATE

The process of removing moisture from an ingredient, in order to dry it out. (See Additional Equipment, page 21, for more information on dehydrating.)

SOUS VIDE

This technique allows you to precisely cook an infusion – alcohol or otherwise – with various ingredients at low temperature while sealed within a vacuum-packed pouch. A sous vide can be as basic as a very large pan filled two-thirds with water and placed over a heat source with a temperature probe to monitor the water temperature. However, sous-vide machines are a much more reliable and less labour intensive option. Cooking under vacuum allows you to heat the liquid to a lower cooking temperature and over a shorter period, thus controlling the process and retaining the integral flavours. Home machines are now widely available.

DISTILLATION AND VACUUM EVAPORATION

The Rotary Evaporator (see Additional Equipment, page 21), is one of the most elaborate pieces of kit I use. It is not a cheap method but produces by far the most insane results. It enables you to distil alcohol or any other liquid with various compounds or ingredients, by way of controlling the pressure. By reducing the pressure, you can control the boiling point by which the infusion evaporates. (For more information on distilling, see page 154.)

STIR

DISTILLATION AND VACUUM EVAPORATION

SOUS VIDE

STRAIN

DRY STORES

The well-prepared bartender needs more than just a good selection of kit and a varied assortment of booze. A well-stocked larder will ensure that you have everything you need to knock up most of the cocktails in this book. Listed below are a list of basics to get you started.

SUGARS

Sugar, in moderation, is no bad thing in a cocktail. Along with bitters (see pages 28–29) and salt, it has the ability to elevate the flavour of your cocktail – as well as the obvious sweetening.

SUGAR SYRUP

The most basic of sweetening modifiers. This is a mix of sugar and hot water, mixed until the sugar is dissolved. Whilst some prefer to use a higher concentration of sugar to water – a 2:1 ratio of sugar to water – I prefer an equal 1:1 ratio and have based all the recipes in this book on the same spec. Make a big batch and store it in the fridge for up to a few weeks or when you need it.

To make a sugar syrup, simply bring equal parts of water and sugar to the boil in a pan, stirring to ensure the sugar is dissolved in the liquid. Set aside to cool, then decant into a bottle and reserve in the fridge until required.

GRENADINE SYRUP

Traditionally made, this is a red syrup produced from pomegranates, though many pre-bottled options include other fruit colourings and additives. Making your own is simple enough and has the added bonus of controlling exactly what it contains. Using the same preferred ratio as the sugar syrup recipe above, simply combine good-quality pomegranate juice and sugar (1:1) over heat until the sugar has dissolved.

HONEY SYRUP

This is a great option for drinks that require a little more depth, rather than just sweetness. I find cutting the honey with water (again, using the same ratio as above) helps with its mixability.

AGAVE

The syrup from the agave plant works great in Tequila and Mezcal cocktails.

MAPLE SYRUP

The sugary sap from the maple tree. Sweeter than regular sugar syrup and 1–3 bar spoons is more than enough to add depth to a drink.

ACIDS

Capable of balancing a drink with only a few drops, acid powders come in various forms and are made from a number of different raw ingredients. Whilst they are great for adding a mouth-puckering sour note to cordials and the odd drop can add a dryer note to a cocktail, be careful when using as it is easy to add too much and completely ruin a drink.

TARTARIC

In most plants, tartaric acid is rare, but it is found in significant concentrations in grapes.

CITRIC ACID

Citric acid exists in greater than trace amounts in a variety of fruits and vegetables, but is most notable in citrus fruits such as lemons, limes, oranges and grapefruits.

MALIC ACID

Malic acid is found in many fruits such as rhubarb and grapes, but is most associated with green apples.

SALTS

As well as being one of the five primary tastes our tongue can detect, salt is a flavour enhancer – it has the ability to suppress bitterness and adds a further dimension to sweet ingredients. When using in drinks, mix it with water to make a salt solution (using 1% salt to water). Adding a drop or two will dramatically change a drink's profile.

EMULSIFIERS

An emulsifier is a substance that stabilizes a mixture of two or more unmixable ingredients by binding them together. Commonly seen in shop-bought produce, such as a vinaigrette (oil in vinegar) or mayonnaise (oil in water). Common emulsifiers include egg and soy lecithin.

HYDROCOLLOIDS

Hydrocolloids are gums that are added to foodstuffs to control their functionality, by way of gelling or thickening. These are most commonly used to make preserves. When you put yesterday's leftover gravy in the fridge overnight – only to find it a wobbling gel, the next morning – this is a hydrocolloid at work. I use hydrocolloids to thicken and clarify ingredients in order to concentrate their flavour.

AGAR AGAR

So good they named it twice. Agar is extracted from algae (seaweed to you and me), which makes it suitable for vegetarians (unlike gelatine, which is derived from collagen). This is used for a variety of results, but for the most part I use this to clarify liquids.

PECTIN

Pectin is a naturally occurring thickening agent that is most often used when added to jams, jellies and similar products to help them gel and thicken. It is found in and around the cell walls of plants and fruits. Pectin performs best when it is heated, So, when making a gel or paint, always ensure the mixture is made when warm. (See Eden, page 64, or Defected Black Russian, page 146.)

BITTERS
(A BARTENDER'S SEASONING)

Originally produced as tonics, remedies and stimulants to cure all manner of ailments, bitters are now a common ingredient in most bars and many companies produce an array of flavours to suit a variety of drinks. These original formulas of herbs, spices, fruits and roots are well-guarded secrets and there are some exquisite flavours on the market that can transform a cocktail with the addition of just a few drops. Whilst there is a wide range of bitters for sale, it is also fun to make your own. The cocktails in this book use both shop-bought and these home-made options:

VANILLA BITTERS

1 vanilla pod, halved and chopped
4 g/0.14 oz vanilla essence
60 ml/2 fl oz vodka

Place the vanilla pod, essence and vodka in a non-reactive container and leave to infuse for 24 hours. Pass the mixture through a coffee filter or muslin-lined sieve (strainer) and collect the liquid. Decant into a pipette bottle and reserve until required.

DRY BITTERS

25 g/¾ oz birch bark*
100 ml/3½ fl oz vodka

Place the birch bark and vodka in a non-reactive container and leave to infuse for 72 hours. Pass the mixture through a coffee filter or muslin-lined sieve (strainer) and collect the liquid. Decant into a pipette bottle and reserve until required.

BURNT CARAMEL BITTERS

100 g/3½ oz caster (superfine) sugar
75 ml/2½ fl oz vodka

Place the sugar in a pan over a medium heat and cook to a dark caramel, being careful not to completely burn the sugar. Remove the pan from the heat and carefully add the vodka. Return the pan to a low heat and stir until the caramel has dissolved into the vodka. Set aside to cool, then decant into a pipette bottle and reserve until required.

CRANBERRY BITTERS

150 g/5½ oz cranberries, roughly chopped
100 ml/3½ fl oz vodka

Place the cranberries and vodka in a non-reactive container, cover and leave to infuse for 24 hours. Pass the mixture through a coffee filter or muslin-lined sieve (strainer) and collect the liquid. Decant into a pipette bottle and reserve until required.

NEROLI BITTERS

8 g/¼ oz dried bitter orange skin*
100 ml/3½ fl oz vodka

Preheat the water bath to 55°C/131°F. Place the ingredients in a vacuum bag and seal. Place the vacuum bag in the water bath and leave to cook for 45 minutes, then remove from the bath and set aside to cool. Strain the infusion through a coffee filter or muslin-lined sieve (strainer), then decant the liquid into a pipette bottle and reserve until required.

BLACK CARDAMOM & PINEAPPLE BITTERS

3 black cardamom pods, lightly crushed
100 g/3½ oz dried pineapple*
150 ml/5 fl oz vodka

Place all the ingredients in a vacuum bag and seal. Leave to infuse for 24 hours, then strain the mixture through a coffee filter or muslin-lined sieve (strainer) and reserve the liquid. Decant into a pipette bottle and reserve until required.

BLOODY BITTERS

20 g/¾ oz green peppercorns
15 g/½ oz dried barbecue spices
2 g /0.07 oz cumin seeds
100 ml/3½ fl oz vodka

Preheat the water bath to 50°C/122°F. Place the ingredients in a vacuum bag and seal. Place the vacuum bag in the water bath and leave to cook for 45 minutes, then remove from the bath and set aside to cool. Strain the infusion through a coffee filter or muslin-lined sieve (strainer), then decant the liquid into a pipette bottle and reserve until required.

CUCUMBER BITTERS

2 g (0.07 oz) food-grade cucumber essence
60 ml/2 fl oz vodka

Mix the ingredients thoroughly and allow to settle, then decant into a pipette bottle and reserve until required.

MANDARIN BITTERS

1 g (0.35 oz) food-grade mandarin essence
60 ml/2 fl oz vodka

Mix the ingredients thoroughly and allow to settle, then decant into a pipette bottle and reserve until required.

*you can pick up dried fruit and fruit skins online or at specialist herb and spice suppliers. See also Suppliers, page 170.

LET'S GET ONE THING CLEAR
(A WORD ABOUT ICE)

Ice is one of the most crucial elements of a drink. It not only chills your liquid to the correct temperature, but also dilutes your drink. And since as much as half the volume of a cocktail can be melted ice, why not pay a little more attention to what you put in your glass? In a stirred drink, the correct ice chills without over-diluting. In a shaken drink, solid cubes add texture as small shards break away from the main mass of the cube, aerating the liquid and giving it mouthfeel.

Whilst your local convenience store will no doubt stock small bags of ready-made ice, these are often already chipped and broken into smaller pieces, their contents consisting of cloudy, often 'wet', ice. Good-quality ice to me is dry, not wet to look at or to touch. It has a distinctive sound when shaken and is as solid as it can possibly be. A crystal-clear shard or block of ice in your drink instantly elevates it to something really special.

For great-quality ice at home, always use water that is either distilled or pre-boiled. Before making your ice, ensure that the water has been chilled in the fridge. This part chilling prior to use will help in allowing any trapped gases, to escape and result in clearer, bubble-free ice.

ICE CUBES

This is the most basic of ice that you will use, but it still deserves your attention. When making your own, search the internet for trays of silicone moulds. Once the ice is frozen, pop it out of the moulds, place in a sealable bag and return to the freezer. This way you can re-use the mould to build up your stock.

ICE BLOCKS/BALLS AND SPHERES

Though larger ice cubes or blocks of ice create greater aeration and texture when shaking and a quicker chill rate than smaller cubes, they do not offer as much dilution as smaller cubes. These are great for drinks such as a Negroni or an Old Fashioned – drinks that require time to melt. Often with this style of drink, I stir the ingredients over regular cubes before transferring to a glass with a large block or sphere.

CRUSHED ICE

Rather than investing in a table-mounted manual ice crusher that produces crappy snow-like ice (great for sushi, though), you can achieve great results by placing some cubes in a kitchen cloth and whacking the cloth with a rolling pin. A good few smacks will give you adequate crushed/chipped ice to make any frappé or swizzle cocktail.

DRY ICE

Dry ice is great for effects such as mist or smoke, both in or around a drink, and is frequently used to preserve foods such as ice creams when being transported. In its gaseous state, dry ice is better known as carbon dioxide. It is most effective as it can achieve a much lower temperature than traditional water-based ice (−78.5 °C or −109.3 °F). I use crushed dry-ice pellets in place of liquid nitrogen to flash-freeze an alcoholic mix into a sorbet or ice cream as I want a higher level of alcohol content but still to retain the traditional sorbet-like consistency.

FLOWERS, FRUIT OR HERB ICE CUBES

These simple and effective cubes look great in a cocktail or G&T. Just make sure that any flowers that you use are edible. Remember that any ingredient you use will be lighter in weight than water, so either prepare your cubes in stages or ensure you pack your moulds with the infusion before adding the water, to ensure that the infused elements do not float to the top. The process is explained step-by-step below:

Place your flowers, fruit or herbs into your moulds (I find silicone moulds give the best results), then one-third fill with cooled water.

Place in the freezer and leave to freeze, then remove and add another third of chilled water. (Add more flowers or fruit at this point, if you feel your ice needs it.)

Return to the freezer and allow to freeze again. Add the remaining third of water and allow to freeze until needed.

INSPIRED BY
THE GARDEN

THAI SWIZZLE

Sweet, spicy, salty and sour, there's nothing like the pungent flavours of Thailand to wake up the senses and get you moving, even on the coldest or most miserable of days. Inspired by one of my favourite Thai dishes, tom yam soup, I created this drink with the vibrant aromas of lemongrass, lime leaf and chilli in mind. Churned through crushed ice, these classic Thai ingredients meld to create a drink that is equal parts fiery, flavoursome and refreshing.

FOR THE COCKTAIL

1 stalk lemongrass, trimmed
4–5 fine red chilli slices
1 lime leaf, finely sliced
1 small pinch micro coriander (cilantro)
50 ml/1¾ fl oz Coriander (Cilantro) Gin (see page 156)
25 ml/¾ fl oz sugar syrup (see page 26)
15 ml/½ fl oz lemon juice

To make the cocktail, finely slice the bottom 1 cm/½ in of the lemongrass stalk and place in the base of the glass along with the sliced chilli, lime leaf and a pinch of micro coriander.

Add the Coriander (Cilantro) Gin, sugar syrup and lemon juice to the glass and one-third fill with crushed ice. Using a bar spoon, churn the ice with the ingredients to meld the flavours, then fill another third of the glass with ice and churn again. Finally, fill the glass to the brim with more ice and garnish with the lemongrass stalk. Serve.

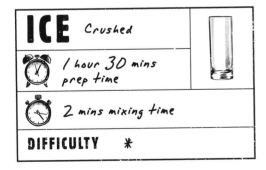

ICE Crushed

1 hour 30 mins prep time

2 mins mixing time

DIFFICULTY *

REMOVED MANDARIN AVIATION

When I worked in Soho, the Aviation was one of a small handful of gin-based cocktails that I would regularly order at my local snacking joint – The Player. I adapted this version as a fragrant, more mouth-enriched version a few years ago. It uses a good old clarified juice, which gives a beautiful translucency and a concentration of flavour and fragrant mandarins.

FOR THE COCKTAIL

60 ml/2 fl oz Hendrick's gin
30 ml/1 fl oz Clarified Grapefruit Juice (see below)
20 ml/¾ fl oz sugar syrup (see page 26)
1 bar spoon maraschino liqueur
8–10 drops Mandarin Bitters (see page 29)
8 drops citric acid

FOR THE CLARIFIED GRAPEFRUIT JUICE

Yields 750 ml/1¾ pints

750 ml/1¾ pints fresh pink grapefruit juice
2 g/0.07 oz agar agar

To make the Clarified Grapefruit Juice, pour 400 ml/14 fl oz of the juice into a pan and place over a medium heat. Heat until just scalding, then gradually pour in the agar agar and stir until completely dissolved. Remove the pan from the heat and pour in the remaining juice. Allow to cool to room temperature, then transfer to a non-reactive container with a lid and place in the freezer, covered, for 24 hours.

The next day, remove the frozen grapefruit juice from the freezer and leave to melt through a muslin-lined sieve (strainer) or coffee filter into a large bowl. Decant the liquid into a bottle and reserve in the fridge until required.

To make the cocktail, place all the ingredients in a shaker or mixing tin, then half-fill with good-quality cube ice. Stir with a bar spoon until icy cold, then double strain into a pre-chilled small Martini glass. Serve.

ICE Cube

24 hours prep time

2 mins mixing time

DIFFICULTY ***

If you can't find good-quality mandarin essence, try the following: peel 260 g/9½ oz mandarins. Add the peel and 350 ml/12 fl oz gin to the sous vide and cook for 45 mins at 48°C/118°F. Cool the infusion and pass through a fine sieve (strainer).

ROASTED RED PEPPER & BLOOD ORANGE BELLINI

Bellinis are a great welcome cocktail for house or garden parties. I love trying to use seasonal produce and playing around with savoury flavours as an alternative to the classic peach. This version contains blood orange juice and red (bell) peppers, which complement each other beautifully as they are both naturally sweet, though the pepper brings a slight vegetable undertone to counter any excessive sweetness. If you prefer a sweeter end result, you can add a little sugar syrup to the finished drink to make it suit your palate.

FOR THE COCKTAIL
65 ml/2¼ fl oz Red Pepper Mix (see below)
65 ml/2¼ fl oz prosecco

FOR THE RED PEPPER MIX
Yields approx. 300 ml/10 fl oz

12 red (bell) peppers, halved and central
   core and ribs removed
225 ml/8 fl oz water
50–75 ml/1¾–2½ fl oz blood orange juice, to taste
sugar syrup (see page 26), to taste (optional)

To make the Red Pepper Mix, preheat the grill (broiler) to high. Lay the peppers, skin-side up, on a baking sheet and place under the grill until the skin is blistered and lightly charred. Transfer the peppers to a resealable plastic bag and set aside for 5–10 minutes to allow the skin to loosen and the peppers to cool.

Once the peppers have cooled, peel off and discard the skin and place the flesh in a blender with 225 ml/ 8 fl oz of water. Blend the mixture until smooth, then pass through a coffee filter or muslin-lined sieve (strainer) to remove any fibrous material. Add the blood orange juice to taste, then sweeten with the sugar syrup if necessary. Decant the liquid into a bottle and reserve in the fridge for up to 72 hours, until required. (If the mixture starts to lose its bright red colour it has been kept too long and should be discarded.)

To make the cocktail, place the Red Pepper Mix and prosecco in a Champagne flute and stir gently to combine. Serve.

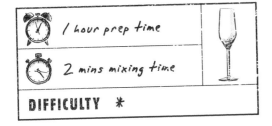

1 hour prep time

2 mins mixing time

DIFFICULTY   ✻

PUMPKIN & CLEMENTINE BELLINI

There are few things that visually signify the onset of the colder months better than the arrival of pumpkins. Children (and adults!) take great pride in carving them with the creepiest of faces and placing them prominently on doorsteps and window ledges to scare unwitting passers-by. But what to do with the discarded innards after you have finished creating your spooky masterpiece? The sweet flesh of vibrantly hued pumpkin makes an excellent vegetable substitution for traditional fruit in this seasonal Bellini. This is a light, vibrant and earthy alternative to the Venetian classic and would be the perfect serve at a Halloween party.

FOR THE COCKTAIL

60 ml/2 fl oz Pumpkin and
    Clementine Juice (see below)
10 ml/0.35 fl oz chestnut liqueur
1 bar spoon Honey Syrup (see
    below)
65 ml/2¼ fl oz prosecco

FOR THE PUMPKIN AND CLEMENTINE JUICE

Yields approx. 300 ml/10 fl oz

1 small pumpkin (approx. 700 g/
    24 oz), peeled and flesh chopped
juice of 8–10 clementines

FOR THE HONEY SYRUP

Yields 100 ml/3½ fl oz

50 ml/1¾ fl oz runny honey
50 ml/1¾ fl oz boiling water

To make the Pumpkin and Clementine Juice, pass the prepared pumpkin flesh through an electric juicer and collect the juice. Strain the juice through a sieve (strainer) into a large measuring jug to remove any debris. Make a note of how much juice you have collected, then add half as much clementine juice as you have pumpkin juice to the jug. Stir to combine, decant into a bottle and store in the fridge until required, or up to 48 hours.

To make the Honey Syrup, combine the honey and water in a small jug or bowl and mix to dissolve the honey. Set aside to cool to room temperature, then store in the fridge until required.

To make the cocktail, place the Pumpkin and Clementine Juice, chestnut liqueur and Honey Syrup in a mixing glass or jug and stir to combine. Pour the mixture into a Champagne flute, pour over the prosecco and stir gently to combine, then serve.

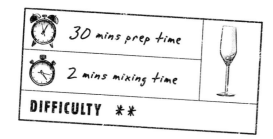

30 mins prep time

2 mins mixing time

DIFFICULTY ✱✱

PINOT PEACH SMASH

Mmmm peaches! Whether they are baked and served with some ice cream, used in a fragrant cocktail or simply eaten raw and fresh so that the juices dribble down your chin, they bring a real burst of sunshine and summer spark to any occasion. This drink is a super simple serve, which makes it perfect for a garden party or picnic with friends. It is also relatively low in alcohol, making it a good choice for an afternoon of drinking that you still want to remember the next morning. The tannins in the tea perfectly complement the peach and mint whilst helping to balance the sweetness.

FOR A SINGLE SERVE

75 ml/2½ fl oz pinot grigio
25 ml/¾ fl oz crème de pêche
2 dashes lemon bitters
1 sprig mint
15 ml/½ fl oz Tea Sugar Syrup (see below)
¼ peach

FOR A LARGE BATCH, SERVING 10-12

750 ml/1¾ pints pinot grigio
250 ml/9 fl oz crème de pêche
20 ml/¾ fl oz lemon bitters
5 sprigs mint
150 ml/5 fl oz Tea Sugar Syrup (see below)
1 peach, cut into wedges

FOR THE TEA SUGAR SYRUP

Yields 250 ml/9 fl oz

250 ml/9 fl oz sugar syrup (see page 26)
4 Earl Grey tea bags

To make the Tea Sugar Syrup, place the sugar syrup in a pan over a medium heat and bring just to the boil. Once boiling, remove the pan from the heat and add the tea bags to the pan. Stir to allow the tea flavour to start to infuse into the syrup, then set aside to cool to room temperature. Once the mixture has cooled, remove and discard the tea bags and decant the infused syrup into a bottle. Store in the fridge until required.

To make the cocktail, add the wine, crème de pêche and lemon bitters to a large wine glass. Take the mint sprig and clap it between your hands to release the aromas, then drop into the glass. Add the Tea Sugar Syrup and peach wedge, then top up the glass with good-quality cube ice. Use a bar spoon to mix until well combined and icy cold. Serve.

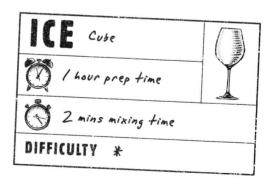

ICE Cube

1 hour prep time

2 mins mixing time

DIFFICULTY *

To make the batched version of this drink, simply add all the ingredients to a punch bowl or large water jug filled with cube ice. Mix to incorporate the flavours, then pour into glasses.

SHISHITO PEPPER CAIPIRINHA

The caipirinha – Brazil's national drink – is a flavour explosion of muddled limes and sugar, spiked with cachaça and churned with crushed ice. I created this drink for SUSHISAMBA's culinary cocktail list in both London and New York. As well as an undercurrent of spice, the peppers inject an earthy note which further enhances the flavour of the cachaça. Around one in every ten of the peppers being spicy, which gives a round of these a fun chilli Russian roulette kinda vibe. Perfect for sharing with friends.

FOR THE COCKTAIL

½ lime, cut into 3 wedges, plus an extra wedge, to garnish
2 brown sugar cubes or 15 ml/½ fl oz Brown Sugar Syrup (see below)
6 medium shishito peppers, grilled (broiled) to a light char and cooled
60 ml/2 fl oz cachaça

FOR THE BROWN SUGAR SYRUP

Yields 300 ml/10 fl oz

150 ml/5 fl oz water
150 g/5½ oz brown sugar

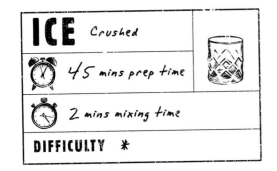

To make the Brown Sugar Syrup, place the water in a pan over a medium heat and bring to a gentle simmer. Add the sugar and stir until dissolved. Remove the pan from the heat and set aside until cooled to room temperature, then decant the liquid into a bottle and reserve in the fridge until required.

To make the cocktail, place the lime wedges in the base of an old fashioned glass and add the sugar or Brown Sugar Syrup, 5 of the grilled (broiled) peppers and half of the cachaça. Use a muddler to release the oils from the lime and peppers and break down the sugar into the cachaça. Add the remaining cachaça to the glass and top up with crushed ice until two-thirds full. Using a bar spoon, churn the drink to incorporate the flavours. Top the glass up with more ice if necessary and serve garnished with another wedge of lime and the remaining pepper.

ICE Crushed
45 mins prep time
2 mins mixing time
DIFFICULTY *

PINE-NEEDLE APEROL SPRITZ

There's nothing more refreshing than a thirst-quenching Aperol spritz. In this version the distinctive aromatic flavour of pine needles offers a playful balance to the richness of the Aperol while the prosecco adds a light, crisp finish. The perfect drink for a summer's evening.

FOR THE COCKTAIL

50 ml/1¾ fl oz Pine-infused Aperol (see page 156)
50 ml/1¾ fl oz soda water
50 ml/1¾ fl oz prosecco
blood orange wedge, to garnish

To make the cocktail, two-thirds fill a large wine glass with good-quality cube ice. Add the Pine-infused Aperol, followed by the soda water and prosecco, then use a bar spoon to stir until well combined and icy cold. Serve garnished with a wedge of blood orange.

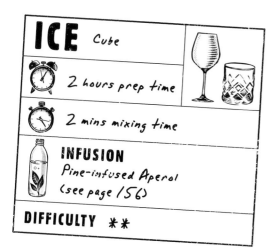

ICE Cube

2 hours prep time

2 mins mixing time

INFUSION
Pine-infused Aperol
(see page 156)

DIFFICULTY ✱✱

If you can't get your hands on any food-grade pine essence, try adding 125g/4½ oz of pine needles when making the infusion. Just remember to wash them before use. Infuse and prepare the same way as described on page 156.

AVOCADO BATIDA

Often made with fruit and cream, the batida is traditionally a working man's drink in its native Brazil. This luxurious dairy-free version contrasts the indulgent buttery texture of avocado with the slight bitterness of chocolate and nutty pistachio syrup for a perfectly balanced result. I created this drink, along with the Shishito Pepper Caipirinha (see page 42) for SUSHISAMBA's culinary cocktail menus.

FOR THE COCKTAIL

50 ml/1¾ fl oz cachaça
50 ml/1¾ fl oz Avocado Purée (see below)
20 ml/¾ fl oz Pistachio Syrup (see below)
1 teaspoon Mozart Chocolate Vodka
mint sprig, to garnish

FOR THE AVOCADO PURÉE

Yields 300 ml/10 fl oz

5 avocados, peeled and flesh chopped
125 ml/4 fl oz water

FOR THE PISTACHIO SYRUP

Yields 500 ml/18 fl oz

125 g/4½ oz shelled pistachio nuts, coarsely chopped
500 ml/18 fl oz sugar syrup (see page 26)

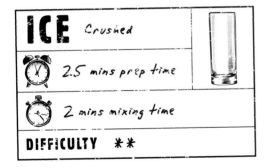

ICE   Crushed

2.5 mins prep time

2 mins mixing time

DIFFICULTY   ✱✱

To make the Avocado Purée, place the flesh of the avocados in a blender with 125 ml/4 fl oz of water and blend until smooth and lump free. Decant the liquid into a squeezy bottle or non-reactive container with a lid and reserve in the fridge until ready to use.

To make the Pistachio Syrup, place the pistachio nuts and sugar syrup in a blender and blend until as smooth as possible, then strain the liquid through a fine sieve (strainer) to remove any debris. Decant the liquid into a bottle and reserve in the fridge until ready to use.

To make the cocktail, fill a large highball glass with crushed ice and add the cachaça, Avocado Purée, Pistachio Syrup and chocolate vodka. Mix the ingredients with a bar spoon until well combined and icy cold. Garnish with a mint sprig and serve.

ORANGE & BASIL FIZZ

Basil and apricot may not seem like usual bedfellows, but try throwing a few basil leaves over a fruit salad and you will see that this unlikely combination can yield magic. This joyful summer sipper is light, fresh and playful, perfect for long sunny days enjoyed with friends.

FOR THE COCKTAIL

8–10 basil leaves, torn
35 ml/1¼ fl oz Bitter Orange Neroli Gin (see page 156)
15 ml/½ fl oz Briottet crème d'abricot
15 ml/½ fl oz sugar syrup (see page 26)
50 ml/1¾ fl oz prosecco
sprig of basil, to garnish

To make the cocktail, place all the basil, Bitter Orange Neroli Gin, crème d'abricot and sugar syrup in a shaker and fill with good-quality cube ice. Shake hard until icy cold, then double strain into a glass filled with cube ice. Add the prosecco and stir gently to combine, then top up with fresh ice if necessary and garnish with a sprig of basil. Serve.

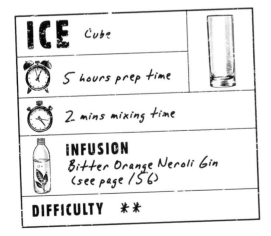

ICE Cube
5 hours prep time
2 mins mixing time
INFUSION Bitter Orange Neroli Gin (see page 156)
DIFFICULTY ＊＊

POLLEN

This light and vibrant cocktail has a wonderful aroma and a taste that is rich in dill but retains a light floral freshness. Sherry is enjoying a quiet renaissance at the moment and makes a great aperitif and accompaniment to pre-dinner snacks and nibbles, such as olives, nuts and bread and oil. I love sherry's distinctive mouth-puckering dryness and when creating this cocktail I wanted to keep as close as possible to its original DNA while elevating its aromatic qualities.

FOR THE COCKTAIL

35 ml/1¼ fl oz Dill Pollen-infused Sherry (see page 156)
15 ml/½ fl oz elderflower cordial
75 ml/2½ fl oz prosecco
1 bar spoon sugar syrup (see page 26, optional)

Place the Dill Pollen-infused Sherry, elderflower cordial and prosecco in a Champagne flute and stir to combine. Taste and add sugar syrup to sweeten, if required. Serve.

1 hour prep time
2 mins mixing time
INFUSION Dill Pollen-infused Sherry (see page 156)
DIFFICULTY ＊＊

Image: Pollen

VIRGIN FENNEL FIZZ

Fennel's aromatic taste is quite unique; it is strikingly reminiscent of liquorice and anise with a slight sweetness that complements the fragrances of the elderflower and dill in this non-alcoholic drink.

FOR THE COCKTAIL

50 ml/1¾ fl oz Fennel Juice (see below)
25 ml/¾ fl oz Dill Sugar Syrup (see below)
15 ml/½ fl oz lemon juice
1 bar spoon elderflower cordial
35 ml/1¼ fl oz soda water
fennel leaves, to garnish

FOR THE FENNEL JUICE

Yields 250 ml/9 fl oz

2 fennel, cleaned and chopped

FOR THE DILL SUGAR SYRUP

Yields 250 ml/9 fl oz

75 g/2¾ oz dill
250 ml/9 fl oz sugar syrup (see page 26)

To make the Fennel Juice, pass the fennel pieces through an electric juicer and collect the juice. Strain the juice through a muslin or tea towel, then decant into a bottle and store in the fridge until required.

To make the Dill Sugar Syrup, preheat the water bath to 45°C/113°F. Place the dill and sugar syrup in a vacuum bag and seal, then place the vacuum bag in the water bath and leave to cook for 45 minutes. Remove from the water bath and set aside to cool, then strain the infusion through a coffee filter or muslin-lined sieve (strainer). Decant into a bottle and store in the fridge until required.

To make the cocktail, shake all the ingredients except the soda water with good-quality cube ice, then double strain into a glass filled with fresh ice cubes. Top up with soda water, mix well to combine and top up the glass with fresh ice, if necessary. Serve garnished with a fennel leaf.

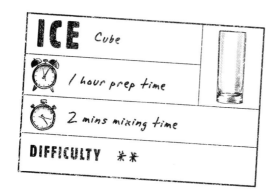

ICE Cube

1 hour prep time

2 mins mixing time

DIFFICULTY ✳✳

For an extra dill hit (and a little liquor), why not add a shot or two of an infused gin. Simply chop 75 g/ 2½ oz dill and cook Sous Vide with 350 ml/12 fl oz gin at 50°C/122°F for 45 mins, before filtering.

BEETROOT & CHOCOLATE ROYALE

If I'm being completely honest, beetroot (beet) isn't one of my favourite vegetables. However, when used as a base flavour for a liqueur or in a cocktail, it imparts a rich earthiness that pairs brilliantly with the creamy bitterness of nuts or chocolate. The liqueur in this cocktail uses this flavour combination to its best advantage and the result is an epic riff on the classic Royale.

FOR THE COCKTAIL

35 ml/1¼ fl oz Beetroot & Chocolate Liqueur
    (see page 157)
90 ml/3 fl oz Champagne

To make the cocktail, pour the liqueur into the base of a Champagne flute and top the glass up with Champagne. Give the drink a gentle stir to combine the ingredients, then serve immediately.

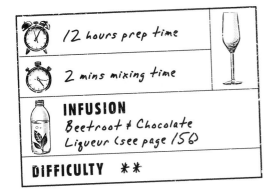

12 hours prep time

2 mins mixing time

INFUSION
Beetroot & Chocolate
Liqueur (see page 156)

DIFFICULTY ✱✱

If you like a more pronounced beetroot (beet) flavour, increase the amount of beetroot to 35 g/1¼ oz when making the liqueur.

LIME – THREE WAYS

Lime brings a refreshing citrus acidity to any drink, but here the flavour is spun out into three distinct and identifiable notes that I find really exciting. The juice and skins bring the familiar zesty notes that we expect from lime, but the lime leaf-infused gin adds a fragrant and subtle savoury backnote that will keep you coming back for more. This is a lovely long and quaffable drink that is great at a bar and even better with friends.

FOR THE COCKTAIL

35 ml/1¼ fl oz Kaffir Lime Leaf Gin (see page 157)
35 ml/1¼ fl oz Lime Husk Cordial (see below)
35 ml/1¼ fl oz egg white
20 ml/¾ fl oz lime juice
soda water, to top up
dried lime leaf powder, to garnish

FOR THE LIME HUSK CORDIAL

Yields 500 ml/18 fl oz

juiced husks of 4 whole limes
500 ml/18 fl oz sugar syrup (see page 26)
1 g/0.035 oz citric acid powder

To make the Lime Husk Cordial, preheat the water bath to 45°C/113°F. Lightly score the skin of the lime husks to release their natural oils, then place them in a vacuum bag with the sugar syrup and seal. Place the vacuum bag in the water bath and leave to cook for 45 minutes, then remove from the water bath and set aside to cool. Strain the infusion through a fine sieve (strainer), mix in the citric acid until dissolved, then decant into a bottle and store in the fridge until required.

To make the cocktail, place all the ingredients except the soda water in a cocktail shaker and dry shake without ice. Fill the shaker with good-quality cube ice and shake again until icy cold. Double strain the mixture into a narrow Collins glass filled with cube ice. Add a splash of soda water and garnish with a sprinkling of dried lime leaf powder.

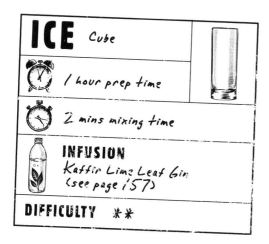

ICE Cube

1 hour prep time

2 mins mixing time

INFUSION
Kaffir Lime Leaf Gin
(see page 157)

DIFFICULTY ✱✱

If you do not have a cream whipper to make the infused gin, simply tear the lime leaves and leave to infuse with the gin in a sealed non-reactive container for a minimum of 4 hours, then filter and reserve.

RHUBARB & DILL SPRITZ

As a child, I wasn't a fan of rhubarb unless it was paired with custard and came in the form of a sugar-coated sweet. Nowadays, I have grown to love the vibrant long-stemmed beauty. I am always excited when it first comes into season and the supermarket shelves are bursting with it. This non-alcoholic cocktail is sweet, sour and refreshing all at once with a slight aniseed tang imparted by the dill.

FOR THE COCKTAIL

Rhubarb and Dill Cordial (see below), to taste
65 ml/2¼ fl oz soda water
fine rhubarb peelings and dill sprigs, to garnish

FOR THE RHUBARB AND DILL CORDIAL

Yields 500 ml/18 fl oz

550 g/1 lb 4 oz rhubarb, cut into 2.5 cm/
   1 in chunks
60 g/2¼ oz dill, roughly chopped
500 ml/18 fl oz water
200 g/7 oz caster (superfine) sugar
2 g/0.07 oz citric acid powder

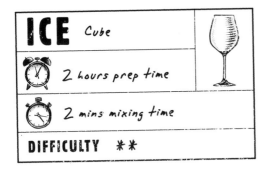

ICE Cube

2 hours prep time

2 mins mixing time

DIFFICULTY ✱✱

To make the Rhubarb and Dill Cordial, preheat the water bath to 50°C/122°F. Place the rhubarb, dill and 500 ml/18 fl oz of water in a vacuum bag and seal. Place the vacuum bag in the water bath and leave to cook for 45 minutes, then remove from the water bath and set aside to cool. Strain the infusion through a muslin-lined sieve (strainer) or coffee filter, then add the sugar and citric acid and stir to combine. Decant the mixture into a bottle and store in the fridge until required.

To make the cocktail, fill a large wine glass with good-quality cube ice, then pour over some of the Rhubarb and Dill Cordial. Top up the glass to two-thirds full with soda water, then taste the drink to check for flavour and sweetness. Add more cordial or soda water as required, then top up with fresh ice. Serve the drink garnished with a fine rhubarb peeling running round the inside of the glass and topped with a sprig of dill.

VERBENA SPRITZ

I love me some sherry. Dry, floral and low in alcohol, it's great as a base ingredient and should not be restricted simply to float on top of a Bloody Mary or gather dust at the back of your booze shelf. By rights, this drink should be a seasonal offering, limiting its appearance to those months where lemon verbena is in season but, if you want to enjoy it year round, you can make the cordial with dried verbena leaves which are available from most good tea specialists.

FOR THE COCKTAIL

25 ml/¾ fl oz Tio Pepe or other dry fino sherry
25 ml/¾ fl oz Lemon Verbena Syrup (see below)
2 dashes lemon bitters
prosecco, to mix
lemon verbena stem, to garnish

FOR THE LEMON VERBENA SYRUP

Yields 1 litre/1¾ pints

35 g/1¼ oz lemon verbena leaves
1 bar spoon citric acid powder
1 litre/1¾ pints sugar syrup (see page 26)

To make the Lemon Verbena Syrup, preheat the water bath to 55°C/131°F. Place the verbena leaves, citric acid powder and 1 litre/1¾ pints of sugar syrup in a vacuum bag and seal. Place the vacuum bag in the water bath and leave to cook for 1 hour, then remove from the water bath and set aside to cool. Strain the infusion through a fine sieve (strainer), then decant the mixture into a bottle and store in the fridge until required.

To make the cocktail, fill a large wine glass two-thirds full with good-quality cube ice. Pour the sherry, Lemon Verbena Syrup and bitters over the ice, then top up with prosecco until the glass is one-third full. Using a bar spoon, stir to combine, then add a few more ice cubes and another third of a glass of prosecco. Stir again, then fill the glass up to the top with the prosecco. Garnish with a lemon verbena stem and serve.

ICE Cube

2 hours prep time

2 mins mixing time

DIFFICULTY ***

Remember to agitate the aroma of the verbena stem slightly by clapping the stem in your hands or slapping it against your palm.

PINE FIZZ

This refreshing drink is great in early spring, when the air is fragrant with blossom. The pine works with the aromatics in the gin and is further elevated by the floral notes of elderflower and lemon.

FOR THE COCKTAIL

50 ml/1¾ fl oz Pine Needle Gin (see page 165)
3 barspoons elderflower cordial
15 ml/½ fl oz sugar syrup (see page 26)
15 ml/½ fl oz Filtered Grapefruit (see below)
1 dash lemon bitters
50 ml/1¾ fl oz soda water
pine sprig, to garnish

FOR THE FILTERED GRAPEFRUIT

300 ml/10 fl oz grapefruit juice

To make the Filtered Grapefruit, pass the grapefruit juice through a muslin-lined sieve (strainer) or coffee filter and collect the juice. Reserve in the fridge until required.

To make the cocktail, place the Pine Needle Gin, elderflower cordial, sugar syrup, Filtered Grapefruit Juice and lemon bitters in a large wine glass and half fill with good-quality cube ice. Stir with a bar spoon to chill the liquids and meld the flavours, then top up with soda and mix again. Top up the glass with more ice if necessary, garnish with a pine sprig and serve.

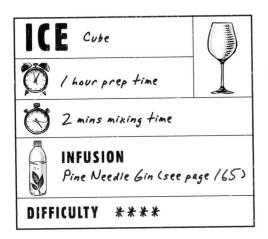

MIND YOUR PEAS & Qs

Flavour and aroma play such an important part in taste, and are a huge trigger for memory recognition. When I was growing up I was fortunate enough to spend many summers evenings eating dinner in the garden. I can still smell the herbs and flowers and the memory is one of my happiest. This simple but epic twist on the gin and tonic is inspired by those days.

FOR THE COCKTAIL

50 ml/1¾ fl oz Pea-infused Gin (see page 157)
75 ml/2½ fl oz tonic water
large mint sprig, to garnish

Fill a highball glass with cube ice, then pour over the Pea-infused Gin. Top up with tonic water and garnish with a large sprig of mint. Serve and enjoy.

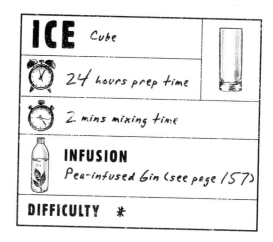

At the point of serving you should inform your guests that this drink is not served with a straw as the aroma of the mint is as important as the flavour of the gin. They should drink from the glass whilst nosing the mint.

CELERY GIMLET

I have spent many hours of blood, sweat and tears trying to perfect this drink. Every time I thought I had finally landed on the perfect recipe, I'd revisit it later and find it needed more work. Any bartenders reading this will understand the frustration of being close to perfection, but never feeling like you have quite nailed it. That being said, everyone else who has tried it has loved it, so I am now turning it over to you to make up your own minds.

Acids have amazing transformative properties and a few drops added to a cocktail can completely change the nature of a drink.

ICE Cube

12 hours prep time

2 mins mixing time

DIFFICULTY ✳✳✳

FOR THE COCKTAIL
35 ml/1¼ fl oz Hendrick's Gin
35 ml/1¼ fl oz Clarified Celery Juice (see below)
4 dashes celery bitters
3 bar spoons sugar syrup (see page 26)
1 bar spoon Citric Acid Solution (see below)
celery peel, to garnish

FOR THE CLARIFIED CELERY JUICE
700 g/1 lb 9 oz trimmed celery sticks
5 g/ ⅛ oz agar agar

FOR THE CITRIC ACID SOLUTION
35 g/1¼ oz citric acid powder
50 ml/2 fl oz vodka

To make the Clarified Celery Juice, pass the celery through an electric juicer and collect the juice (approx. 700 ml/1¼ pints). Pour the juice into a pan and place over a medium heat. Heat until just scalding, then gradually add the agar agar and stir until dissolved. Remove the pan from the heat and allow to cool to room temperature, then transfer to a non-reactive container with a lid and place in the freezer, covered, for 12 hours.

The next day, remove the frozen celery juice from the freezer and leave to melt through a muslin-lined sieve (strainer) or coffee filter into a large bowl. Decant the liquid into a bottle and reserve in the fridge until required.

To make the Citric Acid Solution, combine the acid and vodka and stir until well combined. Decant into a bitters bottle and reserve until required.

To make the cocktail, place all the ingredients in a mixing glass or tin, then two-thirds fill with good-quality cube ice. Stir the drink with a bar spoon until icy cold, then strain into a Martini glass. Garnish with a fine celery peel and serve.

EDEN

In 2014, I entered this drink into the global final of Bombay Sapphire's 'World's Most Imaginative Bartender' competition (WMIB). The idea was to deliver a cocktail that transformed from one style into another. The addition of the IPA, coupled with elderflower brought a bright, floral and hoppy note, whilst the beetroot (beet) paint, once melted into the cocktail, transformed it to an earthy-like, delicate yet intrinsic Martini.

FOR THE COCKTAIL
45 ml/1½ fl oz Bombay Sapphire (or alternative light style gin)
20 ml/¾ fl oz Martini Bianco
20 ml/¾ fl oz IPA Reduction (see below)
1 bar spoon St. Germain Elderflower Liqueur
3 drops Malic Acid Solution (see page 144)
Beetroot (Beet) Paint (see below)

FOR THE IPA REDUCTION
330 ml/11¼ fl oz hoppy IPA
30 g/1 oz caster (superfine) sugar

FOR THE BEETROOT (BEET) PAINT
200 ml/7 fl oz beetroot (beet) juice
2 g/0.07 oz pectin powder
  (approximately 1–1.5% of the juice weight)

To make the IPA Reduction, place the ale in a pan over a medium to low heat and cook until it has reduced by two-thirds (do not attempt to rush this process over a high heat as the ale will lose flavour.) Once reduced, remove from the heat and stir in the sugar. Set aside to cool to room temperature before using.

To make the Beetroot (Beet) Paint, place the beetroot juice in a small pan over a low heat and bring to a gentle simmer. Once simmering, remove the pan from the heat and gradually add the pectin, pulsing the mixture with a stick blender between each addition to ensure the mix is well combined and lump-free.

While still warm, line a sieve (strainer) with a muslin or tea towel and pass the mixture through it to remove any lumps. Transfer the filtered liquid to the fridge to thicken for at least 3 hours before using.

An hour before you want to make your cocktail, place your chosen Martini glasses in the freezer and leave to chill. Once frozen, remove the glasses one at a time and paint a stripe of Beetroot Paint down the side of each. Return the glasses to the freezer before making the cocktail to allow the paint to freeze.

To make the cocktail, place all the ingredients except the Beetroot (beet) Paint in a shaker or stirring jug and fill with good-quality cube ice. Stir until icy cold, then strain into your prepared glass.

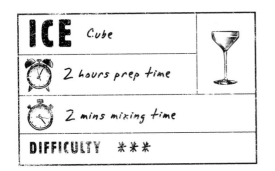

ICE Cube
2 hours prep time
2 mins mixing time
DIFFICULTY ✳✳✳

If the paint has been given enough
time to freeze prior to adding the
cocktail, it should slowly melt into
the drink after a few minutes.

EXPRESSO MARTINI

Most of the people reading this will have enjoyed an Expresso Martini at some point. It was the quintessential 'wake me up and fuck me up' cocktail of the '80s. It was originally created by the late, great Dick Dradsell at the request of a guest and was soon sold in every bar from the beach to the strip. My version contains no coffee, but retains the look, texture and a lot of the flavour of its bigger, brasher brother.

FOR THE COCKTAIL

35 ml/1¼ fl oz vodka
35 ml/1¼ fl oz Expresso Vodka (see page 157)
25 ml/¾ fl oz maple syrup

To make the cocktail, place all the ingredients in a shaker, then dry shake. Fill with good-quality cube ice, then shake again until icy cold. Double strain into a small Martini glass and serve.

You can buy the ingredients needed to make the Expresso Vodka at most herb and spice supply shops or online (see page 170).

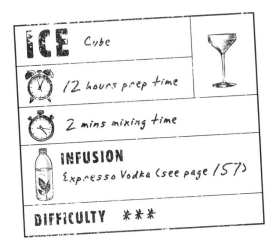

ICE   Cube

12 hours prep time

2 mins mixing time

INFUSION
Expresso Vodka (see page 157)

DIFFICULTY   ✳✳✳

MARIGOLD AFFINITY

The bright orangey-yellow marigold flower has been used for centuries both as a natural remedy to treat many ailments and as a dye. Its flavour brings slightly peppery, honey and hay notes to this drink that, when paired with the sublime St. Germain, creates a flavour that is fragrant, fresh and light.

FOR THE COCKTAIL

45 ml/1½ fl oz Marigold Vodka (see page 158)
20 ml/¾ fl oz sugar syrup (see page 26)
2 bar spoons St. Germain Elderflower Liqueur
1 dash grapefruit bitters
60 ml/2 fl oz prosecco
dried marigold petals, to garnish

Place all the ingredients except the prosecco in a large wine glass, then fill with good-quality cube ice and add a pinch of dried marigold petals. Stir to meld the flavours, then top up the glass with prosecco and additional ice, if needed. Serve.

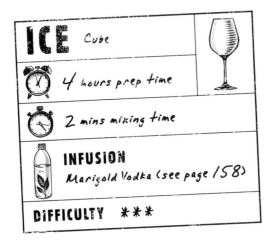

ICE Cube

4 hours prep time

2 mins mixing time

INFUSION
Marigold Vodka (see page 158)

DIFFICULTY ✳✳✳

BLACK CHERRY GODFATHER

The original Godfather was a classic, but for me, it was too sweet. I love the fresh tartness of black cherries and this fruity and drier version makes a wonderful sipping cocktail.

FOR THE COCKTAIL

30 ml/1 fl oz Johnnie Walker Whisky
30 ml/1 fl oz Black Cherry Liqueur (see page 158)
10 drops Burnt Caramel Bitters (see page 28)
10 drops Dry Bitters (see page 28)

To make the cocktail, place a large block of ice into a rocks glass and set aside. Place all the ingredients in a shaker or stirring jug and fill with good-quality cube ice. Stir until icy cold, then strain into your prepared glass. Serve.

FLEUR ROYALE

This is a great welcome drink for a group of drinkers as it is delicious, easy to scale up and simple to make. Other petals and flowers can be used, but I enjoy the rich fragrant aroma of marigold and the sweetness and hue that cornflower gives.

FOR THE COCKTAIL

plum bitters, in an atomizer bottle
30 ml/1 fl oz Liqueur de Fleurs (see page 159)
60 ml/2 fl oz Champagne

To make the cocktail, spray the inside of a large, pre-chilled lare Martini glass with the plum bitters to coat, then add the liqueur. Top up with Champagne and place a petal-filled ice ball (see page 30) in the drink for a final floral touch. Serve.

ICE Block and cube
72 hours prep time
2 mins mixing time
INFUSION Black Cherry Liqueur (see page 158)
DIFFICULTY ****

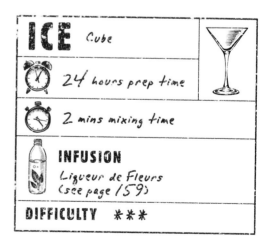

ICE Cube
24 hours prep time
2 mins mixing time
INFUSION Liqueur de Fleurs (see page 159)
DIFFICULTY ***

Image: Fleur Royale

CLARIFIED BLOODY MARY

I love a good Bloody Mary but suffer much like everyone else, as one drink is not enough but two is too many. The filling nature of this beloved brunch-time favourite denies us the pleasure of further consumption. The thinking behind this drink was simple – to create a Bloody Mary that was lighter but equally quaffable and one you could continue to enjoy with food.

FOR THE COCKTAIL

50 ml/1¾ fl oz Grey Goose vodka
2 bar spoons dry vermouth
65 ml/2¼ fl oz Bloody Mary Consommé (see below)
10 drops Bloody Bitters (see page 29)
dehydrated tomato slice, to garnish

FOR THE BLOODY MARY CONSOMMÉ

1 litre/1¾ pints good-quality tomato juice
3 g (0.1 oz) freshly ground black pepper
3 g ((0.1 oz) pink peppercorns, crushed
1 heaped bar spoon sea salt
1 heaped bar spoon celery salt
15 ml/½ fl oz Tabasco sauce
20 ml/¾ fl oz Green Tabasco Sauce
Juice of 1 lemon
50 ml/2 fl oz Worcester Sauce
2-3 bar spoons agar agar

To make the Bloody Mary Consommé, place half the tomato juice in a large pan over a medium heat with the black pepper, pink peppercorns, sea salt, celery salt, both Tabasco sauces, lemon juice and Worcester sauce. Heat to a gentle simmer, then gradually add the agar agar, stirring between each addition to dissolve. Reduce the heat to low and add the remaining tomato juice, then leave to cook and infuse gently for 15 minutes. Set aside to cool, then transfer to a non-reactive container with a lid and transfer to the freezer for at least 12 hours.

The next day, remove the frozen tomato juice from the freezer and leave to melt through a muslin-lined sieve (strainer) or coffee filter into a large bowl. Decant the liquid into a bottle and reserve in the fridge until required.

To make the cocktail, place all the ingredients in a mixing glass or tin, then two-thirds fill with good-quality cube ice. Stir the drink with a bar spoon until icy cold, then strain into a small coupette or Martini glass. Garnish with a dehydrated tomato slice and serve.

ICE Cube

12 hours prep time

2 mins mixing time

DIFFICULTY ✳✳✳✳

ELECTRIC CORIANDER COLADA

Emotional connections to memories tied with drinking or a particular drink are so powerful. The circumstance for me has always been what has made the Piña Colada a disco classic and such a great drink – memories of holidaying on the beach, or with friends. This is when this drink is at its best. That said, there are some great versions out there and this is just one of them. If you are fortunate enough to have a carbonated shaker, omit the soda in this recipe and go right ahead and shake yourself up a fizzy Piña. If not, don't worry. The addition of just a little soda will give you the same effect.

FOR THE COCKTAIL

35 ml/1¼ fl oz Coconut-washed Light Rum (see page 158)
65 ml/2¼ Clarified Pineapple Sherbet (see below)
1 pipette citric acid
1 pinch micro coriander (cilantro), to garnish
soda water

FOR THE CLARIFIED PINEAPPLE SHERBET

1 litre/1¾ pints good-quality pineapple juice
5 g/⅛ oz agar agar
caster (superfine) sugar

2 g/0.007 oz citric acid powder
To make the Clarified Pineapple Sherbet, place 700 ml/1¼ pints of the pineapple juice in a pan and place over a medium heat. Bring just to a simmer, then add the agar agar and stir until dissolved. Remove from the heat and set aside to cool for 15 minutes, then transfer to a non-reactive container with a lid and place in the freezer for at least 14 hours.

The next day, remove the frozen pineapple juice from the freezer and leave to melt through a muslin-lined sieve (strainer) or coffee filter into a large bowl. Measure the collected liquid, then add 100 g/3½ oz of caster (superfine sugar) for every 500 ml/18 fl oz of clarified juice and the citric acid powder. Mix until the sugar and acid have dissolved in the juice, then decant into a bottle and reserve in the fridge until required.

Before making your cocktail, first prepare your glass. Fill a Collins glass with good-quality cube ice and pour approximately 25 ml/¾ fl oz soda water over the ice.

Now prepare the cocktail by adding all the remaining ingredients to a shaker. Fill with good-quality cube ice. Shake hard, then double strain over the soda into the glass and top up with a splash of fresh soda. Garnish with a pinch of micro coriander (cilantro) and serve.

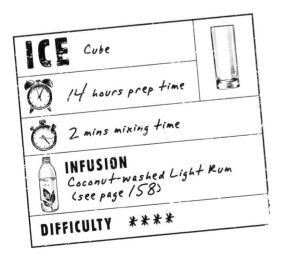

ICE Cube

14 hours prep time

2 mins mixing time

INFUSION Coconut-washed Light Rum (see page 158)

DIFFICULTY ★★★★

HOPPED SCOTCH

I chose Goldings hops for this drink as I wanted a citrus-hoppy like note. You can work with other similar hops, too. Both Cascade and Citra have similar notes with the latter known for its grapefruit- and tropical fruit-like characteristics.

FOR THE COCKTAIL

1 egg white
35 ml/1¼ fl oz Hopped Scotch (see page 158)
25 ml/¾ fl oz sugar syrup (see page 26)
25 ml/¾ fl oz grapefruit juice
2 bar spoons elderflower cordial
1 dash grapefruit bitters
black cardamom pods, to garnish

Place the egg white in a shaker followed by the rest of the ingredients, then dry shake. Fill the shaker with good-quality cube ice and shake again. Double strain into a rocks glass over fresh cube ice. Serve garnished with grated cardamom.

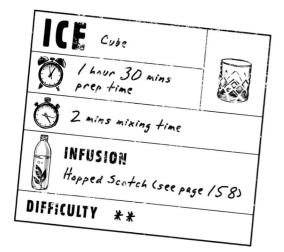

ICE Cube

1 hour 30 mins prep time

2 mins mixing time

INFUSION
Hopped Scotch (see page 158)

DIFFICULTY **

PEA & MINT FIZZY DAIQUIRI

This is a delightfully light and refreshing serve that is a perfect welcome drink on a sunny day. Pea and mint make excellent bedfellows and are beautifully fragrant and reminiscent of the spring and summer months when the garden is in full bloom.

FOR THE COCKTAIL

8–10 sugar snap peas
50 ml light rum (such as Bacardi Carta Blanca)
25 ml/¾ fl oz lime juice
25 ml/¾ fl oz sugar syrup (see page 26)
25 ml/¾ fl oz prosecco
1 small mint sprig, to garnish

Snap the pea pods and drop in to your shaker. Add the rest of the ingredients to the shaker and muddle to break up the pods and infuse with the liquid. Fill the shaker with good-quality cube ice and shake hard for approximately 10 seconds. Double strain into your desired glass and spritz with the prosecco. Garnish with a small mint leaf, floating on the surface of the drink.

ICE Cube

2 mins prep time

2 mins mixing time

DIFFICULTY *

VIRGIN PIÑA-KALE-ADA

Who doesn't like a Piña Colada?! Some of the worst I have ever had have ironically been some of the best – sat atop a cliff in the Caribbean or slumped on the beach. Made well, these are gloriously simple drinks that deserve more than a little appreciation.

This non-alcoholic version is as good as its grown-up sibling, but fear not, those of you wanting more of a kick should simply add 35 ml/1¼ fl oz Coconut-washed Rum (see page 158) to the recipe below.

FOR THE COCKTAIL
65 ml/2¼ fl oz pineapple juice
25 ml/¾ fl oz coconut cream
15 ml/½ fl oz sugar syrup (see page 26)
15 ml/½ fl oz lime juice
15 ml/½ fl oz Kale Juice (see below)

FOR THE KALE JUICE
250 g/9 oz kale juice
125 ml/4 fl oz water

To make the Kale Juice, place the kale and water in a blender and pulse until combined. Pass through a sieve (strainer) and reserve until required.

To make the cocktail, place all the ingredients in a shaker. Fill with good-quality cube ice. Shake hard, then double strain into the glass filled with more good-quality cube ice.

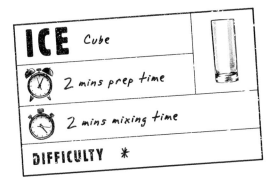

GARDEN SPRITZ

Here, red (bell) pepper is added to give the drink a sweet earthiness that plays brilliantly against the herby bitterness of the Campari and Aperol. I created this drink when it was cold and wet outside, but I was definitely dreaming about long summer evenings.

FOR THE COCKTAIL
30 ml/1 fl oz Campari
30 ml/1 fl oz Aperol
45 ml/1½ fl oz Red Pepper Juice (see below)
30 ml/1 fl oz blood orange juice
soda water, to top up
blood orange wedge, to garnish

FOR THE RED PEPPER JUICE
Yields 200 ml/7 fl oz

5 red peppers, halved and central core and ribs removed

To make the Red Pepper Juice, pass the prepared red (bell) pepper through an electric juicer and collect the juice. Strain the juice through a muslin or coffee filter to remove any debris and reserve the liquid in the fridge until required.

To make the cocktail, fill a Collins glass with good-quality cube ice and add the ingredients, topping the glass up with enough soda water to fill to the top. Mix well with a bar spoon to combine the flavours and garnish with a blood orange wedge. Serve immediately.

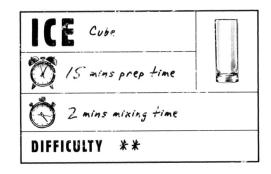

FROM THE
KITCHEN

MULLED CIDER

A piping hot cocktail served round a fire or bonfire on a cold winter's evening can warm the insides and lift the spirit. This mulled cider cocktail is best served in an insulated cup or mug to help preserve the temperature, especially if being served outside. You can play with the blend of flavour-enriching aromatics when making this drink, but I love the combination of ginger and cinnamon as it goes wonderfully with the warming cognac.

FOR THE COCKTAIL

50 ml/1¾ fl oz cognac
3 bar spoons maple syrup
65 ml/2¼ fl oz hot Mulled Cider (see below)
apple slice and cinnamon stick, to garnish

FOR THE MULLED CIDER

Yields 250 ml/9 fl oz

500 ml/18 fl oz pear cider
25 g/¾ oz root ginger, sliced
1 cinnamon stick
zest of ½ an orange

To make the Mulled Cider, place the cider and sliced ginger in pan over a medium heat, bring to a gentle simmer and leave to cook for 5 minutes. Add the cinnamon stick and orange zest and continue to simmer until the liquid has reduced by half. Remove from the heat and strain through a sieve (strainer). Keep warm until required.

To make the cocktail, pour the cognac and maple syrup into the base of a mug or heatproof glass, then pour over the Mulled Cider. Garnish with a slice of apple and cinnamon stick and serve hot.

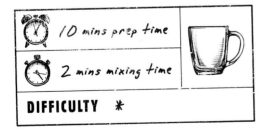

10 mins prep time

2 mins mixing time

DIFFICULTY *

MULLED NEGRONI

All in favour of making the Negroni a must list cocktail on ALL menus, worldwide? Raise your hand! I don't think I've ever written a cocktail menu that doesn't list a Negroni. From the earliest 'Bottled' version and the foraged 'Woodland' variety to the juxtaposed digestif-aperitivo style 'Ristretto' Negroni to the über-decadent Truffled Negroni (see opposite for those last two). So, when I was thinking of warm cocktails, it was given that a hot version of this Italian giant was going to be included. The trick here is not to overcook the alcohol and risk evaporating it. See image page 85.

FOR THE COCKTAIL

150 ml/ 5 fl oz Mulled Negroni Pre-Mix or approximately half your chosen mug (see below)
coins of orange zest and pitted green olives, to garnish

FOR THE PRE-MIX

Yields 1 litre/1¾ pints

250 ml/9 fl oz over proof gin
200 ml/7 fl oz sweet vermouth
200 ml/7 fl oz Campari
100 ml/3½ fl oz maple syrup
250 ml/9 fl oz water
4 bay leaves
1 cinnamon stick, broken
1 g (0.04 oz) white peppercorns
1 tonka bean, broken
50 g/1¾ oz pitted green olives

To make the Pre-Mix, preheat the water bath to 45°C/113°F. Place all the ingredients in a vacuum bag and seal, then place the vacuum bag in the water bath and leave to cook for 50 minutes. Remove from the water bath and set aside to cool, then strain the infusion through a coffee filter or muslin-lined sieve (strainer). Decant into a bottle and store in the fridge until required.

To make the cocktail, pour the Pre-Mix into a large pan and warm gently until it just comes to a simmer (anything more than this will start to cook off the alcohol). Ladle the drink into mugs and garnish each with a coin of orange zest and a pitted green olive.

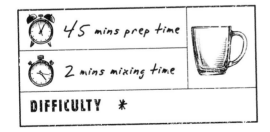

45 mins prep time

2 mins mixing time

DIFFICULTY ✳

RISTRETTO NEGRONI

I created this Negroni variant a few years ago as I wanted people to enjoy this great drink after a meal and not limit its appeal to that of a classic aperitif. The use of coffee adds layers of complexity and further bitterness but also the added aroma of roasted coffee. See image page 84.

FOR THE COCKTAIL
75 ml/2½ fl oz Ristretto Negroni Pre-Mix (see below)
dehydrated orange wheel, to garnish

FOR THE PRE-MIX
300 ml/10 fl oz gin
200 ml/6½ fl oz Campari
125 ml/4 fl oz Aperol
200 ml/6½ fl oz sweet vermouth
25 ml/¾ fl oz sugar syrup (see page 26)
35 g/1¼ oz fresh-roasted coffee beans

To make the Pre-Mix, place all the ingredients in a non-reactive container and leave to infuse for 72 hours, agitating the container every so often. Once the mixture has infused, strain the liquid through a muslin-lined sieve (strainer) or coffee filter and collect the liquid. Decant into a bottle and store in the fridge until required.

To make the cocktail, place the Pre-Mix in a shaker or mixing tin, then fill with good-quality cube ice. Stir for 30 seconds until icy cold, then strain over a block of ice into the glass. Garnish with the dehydrated orange wheel and serve.

TRUFFLED NEGRONI

Along with the other Negroni recipes in this book, this was created for one of the menus at Duck & Waffle. This Fall/Winter 2015 version called for the lovely pairing of black truffle and dark chocolate. An über-luxe combo which perfectly complement one another. See image page 84.

FOR THE COCKTAIL
50 ml/2 fl oz Truffled Negroni Pre-Mix

FOR THE PRE-MIX
250 ml/9 fl oz gin
200 ml/7 fl oz sweet vermouth
200 ml/7 fl oz Campari
25 ml/¾ fl oz Mozart Dark Chocolate Liqueur
25 ml/¾ fl oz sugar syrup (see page 26)
10 dashes chocolate bitters
20 g/¾ oz sliced black truffle

To make the Pre-Mix, place all the ingredients in a non-reactive container and leave to infuse for 72 hours, agitating the container every so often. Once the mixture has infused, strain the liquid through a muslin-lined sieve (strainer) or coffee filter and collect the liquid. Decant into a bottle and store in the fridge until required.

To make the cocktail, place the Pre-Mix in a shaker or mixing tin, then fill with good-quality cube ice. Stir for 30 seconds until icy cold, then strain over a block of ice into your desired glass and serve.

Image (from left to right):
Truffled Negroni (page 83),
Ristretto Negroni (page 83)
and Mulled Negroni (page 82).

CYNAR COKE FLOAT

I'm not sure if this is a cocktail or a dessert and, to be honest, I don't care! This twisted American classic, bittered by the addition of Cynar, is glorious. You can buy cola syrup online or from a soda stream vendor.

FOR THE COCKTAIL
Cynar Vanilla Ice Cream (see below)
50 ml/1¾ fl oz Cynar
25 ml/Vanilla Cola Syrup (see below)
100 ml/3½ fl oz cream soda

FOR THE VANILLA COLA SYRUP
Yields 150 ml/5 fl oz

150 ml/5 fl oz cola syrup
1 vanilla pod, split lengthways

FOR THE CYNAR VANILLA ICE CREAM
Yields 500 ml/18 fl oz

1 x 500 ml/18 fl oz tub vanilla ice cream
75 ml/5 fl oz Cynar

If you are not a fan of cream soda, you can substitute regular cola and still achieve great results.

To make the Vanilla Cola Syrup, place the cola syrup and vanilla pod in a non-reactive container with a lid, cover and leave overnight to infuse. The next day, remove the vanilla pod and decant the syrup into a bottle. Reserve until required.

To make the Cynar Vanilla Ice Cream, remove the ice cream from the freezer and set aside to soften until just mixable. Meanwhile, place the Cynar in a pan over a medium heat and bring to a simmer. Leave to cook until reduced by a third, then remove from the heat and allow to cool to room temperature. Once the Cynar reduction has cooled and the ice cream is softened, pour the syrup into the tub with the ice cream and stir through to create a ripple effect. Return to the freezer to harden and reserve until required.

To make the cocktail, two-thirds fill a sundae glass with Cynar Vanilla Ice Cream, then pour over the Cynar and Vanilla Cola Syrup and top with cream soda. Enjoy with a spoon and a straw.

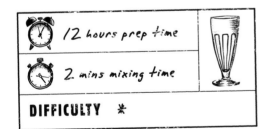

12 hours prep time

2 mins mixing time

DIFFICULTY ✳

STEALTH MARG

Avoid the slushy versions of this drink and try this fiery one instead. The jalapeño peppers add a glorious heat to this Tommy's Margarita spin-off, whilst the St. Germain adds a light and floral note that subdues the peppery flavours.

FOR THE COCKTAIL

60 ml/2 fl oz Jalapeño Tequila (see page 159)
15 ml/½ fl oz lime juice
2 bar spoons St. Germain Elderflower Liqueur
1 bar spoon agave nectar
lime wedges, dried coriander (cilantro) and flaked sea salt, to garnish

To prepare your glass, rub a cut lime wedge around the rim of the glass, then dip into a bowl of dried coriander (cilantro) and sea salt flakes. Set aside.

To make the cocktail, place all the ingredients in a shaker, then stir with a bar spoon to ensure the agave syrup is thoroughly mixed in. Fill with good-quality cube ice, then shake hard until icy cold. Double strain the drink into your prepared glass.

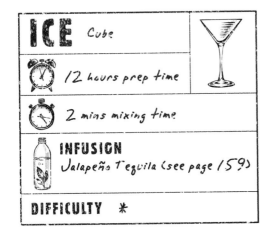

ICE Cube

12 hours prep time

2 mins mixing time

INFUSION
Jalapeño Tequila (see page 159)

DIFFICULTY ✳

If a straight-up Margarita is not your thing, why not shake this and serve it long over cube ice into a salt-rimmed glass. Top it up with grapefruit soda and garnish with a wedge of pink grapefruit.

CHOCOLATE FLIP

This luxurious and decadent cocktail is fragrant with cardamom and chocolate, which perfectly complements the spicy notes of Chartreuse.

FOR THE COCKTAIL

1 egg
45 ml/1½ fl oz green Chartreuse
30 ml/1 fl oz Chocolate and Cardamom Syrup (see below)
2 dashes chocolate bitters
1 dash Cynar
drinking chocolate, to garnish

FOR THE CHOCOLATE AND CARDAMOM SYRUP

Yields 200 ml/7 fl oz

200 ml/7 fl oz sugar syrup (see page 26)
100 g/3½ oz powdered drinking chocolate
3 black cardamom pods, cracked in a pestle and mortar

To make the Chocolate and Cardamom Syrup, place the sugar syrup and drinking chocolate in a pan over a medium heat. As the syrup heats, whisk the chocolate into the liquid until fully incorporated, then remove from the heat and add the cracked cardamom pods. Set aside to infuse and cool to room temperature.

Once cooled, strain the syrup through a fine sieve (strainer) and reserve the liquid. Decant into a bottle, transfer to the fridge and reserve until required.

To make the cocktail, crack the egg into a shaker, followed by the other ingredients. Fill with good-quality cube ice, shake hard until icy cold, then double strain into your glass. Garnish the drink with a light dusting of drinking chocolate and serve.

A shot of the chocolate and cardamom syrup used in this recipe would make an excellent addition to your morning coffee.

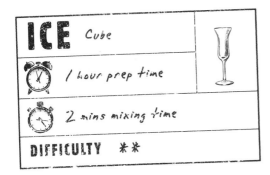

ICE Cube

1 hour prep time

2 mins mixing time

DIFFICULTY ✳✳

PANETTONE FLIP

This is the ultimate in decadence and should be served in a short tasting glass and enjoyed slowly due to its luxurious nature. The addition of panettone lends notes of citrus and dried raisins, whilst the tonka bean brings festive notes of spice and warmth.

FOR THE COCKTAIL

1 egg
45 ml/1½ fl oz Panettone-infused Cognac (see page 159)
25 ml/¾ fl oz Spiced Sugar Syrup (see below)
1 dash orange bitters
freshly grated nutmeg, to garnish

FOR THE SPICED SUGAR SYRUP

Yields 250 ml/9 fl oz

4 tonka beans, broken
½ cinnamon stick
250 ml/9 fl oz sugar syrup (see page 26)

To make the Spiced Sugar Syrup, place the tonka beans and cinnamon stick in a non-reactive container with a lid and pour over the sugar syrup. Cover and leave to infuse overnight. The next day, strain the syrup through a fine sieve (strainer) and collect the liquid. Decant into a bottle and reserve in the fridge until required.

To make the cocktail, crack the egg into a shaker, then add the remaining ingredients. Dry shake, then fill with good-quality cube ice. Shake hard until icy cold, then double strain into a glass. Garnish with a dusting of freshly grated nutmeg and serve.

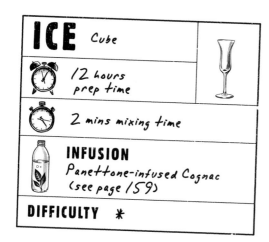

ICE Cube

12 hours prep time

2 mins mixing time

INFUSION
Panettone-infused Cognac
(see page 159)

DIFFICULTY ✳

COSMO DE PROVENCE

Made well a classic Cosmopolitan is a good drink – not my kind of drink, but one I can appreciate nonetheless. This version is a different story and I would never pass one up. It uses the lovely blend of savory, marjoram, rosemary, thyme and oregano which make up herbes de Provence, infused with citron vodka and orange liqueur, all of which result in a beautiful marriage of flavours.

FOR THE COCKTAIL

50 ml/1¾ fl oz Cosmo de Provence Mix (see page 159)
25 ml/¾ fl oz cranberry juice
2 bar spoons lime juice
2 bar spoons sugar syrup (see page 26)
4 drops Cranberry Bitters (see page 28)
coin of orange zest, to garnish

Place all the ingredients in a shaker, then fill with good-quality cube ice. Shake well until icy cold, then double strain into the glass. Squeeze a coin of orange zest to spray the oils over the drink and stem of the glass, then place on the surface of the drink and serve.

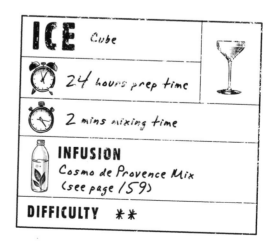

ICE Cube

24 hours prep time

2 mins mixing time

INFUSION
Cosmo de Provence Mix
(see page 159)

DIFFICULTY ✳✳

Image (from top to bottom):
Cosmo de Provence (opposite),
Roasted Cosmo (page 94).

ROASTED COSMO

Let's face it, we all have a guilty pleasure drink. Now, I'm not saying the Cosmopolitan is mine, but that was the inspiration for this drink: to take a cocktail with a stigma attached and tweak it to become a drink that everyone can enjoy.

Of all the drinks I've created, this is one of those that people always ask about – 'why bone marrow?' When I first added this drink to Duck & Waffle's menu back in 2012, I knew I wanted to use the cranberry in this drink and when looking at flavours to pair with it I kept coming back to the idea of roasted meats. Then it was just a case of landing on the ingredient that worked best, which is how I settled on bone marrow. See image page 93. (Fat washing is also used in the Kobe Cocktail page 120 , Chocolate & Blue Cheese Martini page 128 and Bulletproof Old Fashioned page 133.)

FOR THE COCKTAIL

35 ml/1¼ fl oz Roasted Cosmopolitan Mix
 (see page 160)
35 ml/1¼ fl oz cranberry juice
1 bar spoon lime juice
1 bar spoon sugar syrup (see page 26)
1 bar spoon Briottet white crème de cacao
1 rosemary sprig, snapped to release aromas

Place all the ingredients, including the rosemary sprig, in a shaker or mixing tin, then half-fill with good-quality cube ice. Stir with a bar spoon until icy cold, then double strain into a pre-chilled small Martini glass. Serve.

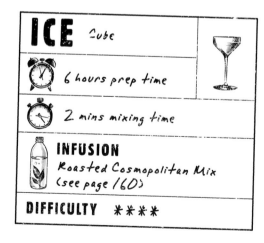

ICE Cube

6 hours prep time

2 mins mixing time

INFUSION
Roasted Cosmopolitan Mix
(see page 160)

DIFFICULTY ✳✳✳✳

DAILY GRIND

I first created this cocktail at Duck & Waffle back in the summer of 2016. The concept of the menu was 'Urban Decay' and the idea was to highlight what flavour could be extracted from ingredients that were considered to be past their best. In the case of this drink, I used spent coffee grinds to infuse an aperitif wine. Using spent coffee grinds not only gives them another purpose, but also allows us to extract a recognizable coffee flavour without it being as overpowering as using fresh beans.

FOR THE COCKTAIL

25 ml/¾ fl oz vodka
25 ml/¾ fl oz Coffee Cocchi (see page 160)
1 bar spoon Yogurt Syrup (see below)
15 ml/½ fl oz Briottet crème d'abricot
2 bar spoons maple syrup
mint sprig and spent coffee grinds, to garnish

FOR THE YOGURT SYRUP

Yields 125 ml/4 fl oz

50g/1¾ oz yogurt powder
100 ml/3½ fl oz water

To make the Yogurt Syrup, place the yogurt powder and water in a bowl and blend until thoroughly combined, using a stick (immersion) blender. Filter the mixture through a fine sieve (strainer) and reserve the liquid. Decant into a small bottle and reserve in the fridge until required.

To make the cocktail, place all the ingredients in a large highball glass and one-third fill with crushed ice. Using a bar spoon, churn the ice with the ingredients to meld the flavours, then fill another third of the glass with ice and churn again. Finally, fill the glass to the brim with more ice and garnish with a mint sprig and a dusting of coffee grinds. Serve.

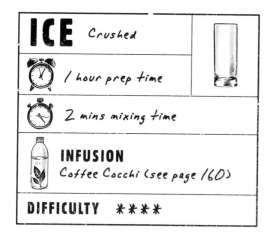

ICE Crushed

1 hour prep time

2 mins mixing time

INFUSION
Coffee Cocchi (see page 160)

DIFFICULTY ★★★★

AUTUMN SANGRIA

Autumn (or fall, depending where you are in the world) is one of my favourite times of year. The leaves are starting to turn, the first nip of cold is in the air and the festive season is on the horizon. This is a great drink to enjoy among friends at this special time of year, when there is a chill in the air but the sun is still bright in the sky.

FOR THE COCKTAIL

350 ml/12 fl oz Seasonal Cognac (see page 161)
500 ml/18 fl oz dry white wine
250 ml/9 fl oz Cinnamon Sugar (see below)
pear slices and blackberries, to garnish (optional)

FOR THE CINNAMON SUGAR

Yields 250 ml/9 fl oz

2 cinnamon sticks, snapped into shards
250 ml/9 fl oz sugar syrup (see page 26)

To make the Cinnamon Sugar, place the cinnamon sticks and sugar syrup in a non-reactive container with a lid. Seal the container and leave the mixture to infuse at room temperature overnight.

Once the mixture has infused, strain into a bottle and reserve in the fridge until needed.

To make the cocktail, fill a large jug or pitcher with good-quality cube ice and add the cocktail ingredients. Stir well to chill and dilute, then set aside for 5 minutes. Strain the drink into serving glasses and serve garnished with sliced pear and blackberries.

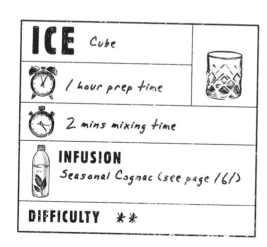

ICE Cube

1 hour prep time

2 mins mixing time

INFUSION
Seasonal Cognac (see page 161)

DIFFICULTY ✳✳

Unlike most of the recipes in this book which are single serves, this makes a large jug that will easily serve 6 people, making it perfect for a party.

MILLENNIAL MAI TAI

Who doesn't love a cheeky Tiki cocktail? This Mai Tai substitutes the traditional orgeat for trendy quinoa. With the addition of coffee and neroli bitters, this is a drier but very flavourful riff on the classic Polynesian cocktail.

FOR THE COCKTAIL

30 ml/1 fl oz Coffee-infused Dark Rum
   (see page 161)
30 ml/1 fl oz cachaça
30 ml/1 fl oz Quinoa Syrup (see below)
15 ml/½ fl oz lime juice
1 dash Neroli Bitters (see page 29)
charred lime husk and mint sprig,
   to garnish

FOR THE QUINOA SYRUP

Yields 400 ml/14 fl oz

150 g/5½ oz red quinoa
350 ml/12 fl oz water
caster (superfine) sugar

To make the Quinoa Syrup, preheat the water bath to 45°C/113°F. Place the quinoa in a frying pan (skillet) over a medium heat and lightly toast. Place the toasted quinoa and the water in a vacuum bag and seal. Place the vacuum bag in the water bath and leave to cook for 50 minutes, then remove from the water bath and set aside to cool. Strain the infusion through a fine sieve (strainer), then weigh the collected liquid and add half the weight again of caster (superfine) sugar. Stir the mixture until the sugar has dissolved, then decant the liquid into a bottle and reserve in the fridge until required.

To make the cocktail, place all the ingredients in a tall glass and two-thirds fill with crushed ice. Using a bar spoon, churn the ice with the ingredients to meld the flavours, then top up the glass with more crushed ice. Garnish with a charred lime husk and a sprig of mint and serve.

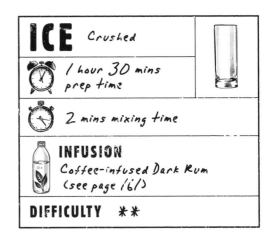

ICE Crushed

1 hour 30 mins prep time

2 mins mixing time

INFUSION
Coffee-infused Dark Rum
(see page 161)

DIFFICULTY ✳✳

SCOTCH COFFEE

Inspired by Irish Coffee, this adds homemade coffee liqueur and a nutty note of praline, making for a wonderful winter warmer. Use an Irish coffee glass if possible.

FOR THE COCKTAIL

30 ml/1 fl oz Scotch whisky
30 ml/1 fl oz Coffee Liqueur (see page 160)
1 bar spoon maple syrup
2 dashes chocolate bitters
90 ml/3 fl oz hot coffee
50 ml/1¾ fl oz Praline Cream (see below)
powdered chocolate, to garnish

FOR THE PRALINE CREAM

250 ml/9 fl oz double cream
100 ml/3½ fl oz sugar syrup (see page 26)
5 g/⅛ oz food-grade praline essence

To make the Praline Cream, place the ingredients in a bowl and lightly hand whip with a balloon whisk. Set aside.

To make the cocktail, place all the ingredients except the Praline Cream in your glass, allowing enough space for the Praline Cream to follow, and gently stir to combine. Spoon over the Praline Cream to cover and garnish with a dusting of powdered chocolate.

WRAY-BENA

Possibly an unlikely combo, but those of you who have ever had hot Ribena will understand. This is a great antidote for a cold, with a warming blackberry taste and a shot of Wray to send all bugs running for the hills. To be honest, I can't remember if this truly worked when I was suffering, but the hit of overproof rum certainly destroyed any chance of me feeling sorry for myself. This cocktail is based on the original cure-all but in a swizzle form. The crushed ice helps to dilute the rum slightly and mellow the flavours.

FOR THE COCKTAIL

35 ml/1¼ fl oz Blackcurrant Wray & Nephew (see page 160)
25 ml/¾ fl oz sugar syrup (see page 26)
½ bar spoon citric acid solution
1 drop lemon bitters
10 ml/0.35 fl oz crème de cassis
blackcurrant or lemon leaf, to garnish

To make the cocktail, place the Blackcurrant Wray & Nephew, sugar syrup, citric acid solution and lemon bitters in a highball glass and one-third fill with crushed ice. Using a bar spoon, churn the ice with the ingredients to meld the flavours, then fill another third of the glass with ice and churn again. Finally, fill the glass to the brim with more ice and garnish with a blackcurrant or lemon leaf. Drizzle the crème de cassis over the top of the cocktail and serve.

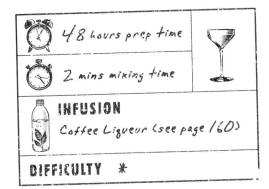

48 hours prep time
2 mins mixing time
INFUSION Coffee Liqueur (see page 160)
DIFFICULTY *

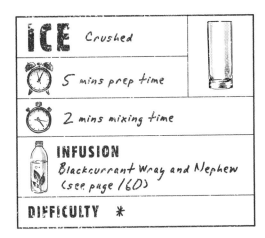

ICE Crushed
5 mins prep time
2 mins mixing time
INFUSION Blackcurrant Wray and Nephew (see page 160)
DIFFICULTY *

MILKYBAR SNOWBALL

This is a modern riff on an '80s classic that uses everyone's sweet-shop fave, chocolate buttons! I make a slightly more technical version of this at the bar, but this is a great home version and the results will have any chocoholic guests coming back for more.

FOR THE COCKTAIL

30 ml/1 fl oz prosecco
1 egg
65 ml/2¼ fl oz Milkybar Liqueur (see page 161)
25 ml/¾ fl oz vodka
15 ml/½ fl oz sugar syrup (see page 26)

To make the cocktail, pour the prosecco into the base of your milk bottle or glass and set aside. Place the remaining ingredients into a shaker and dry shake. Fill the shaker with good-quality cube ice, then shake hard until icy cold. Gently strain into your bottle or glass over the top of the prosecco. Serve.

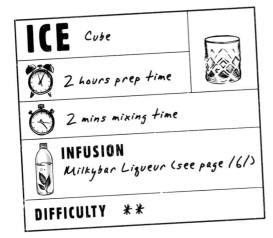

ICE Cube

2 hours prep time

2 mins mixing time

INFUSION
Milkybar Liqueur (see page 161)

DIFFICULTY ✳✳

I like to serve this in novelty milk bottles, though if you can't get hold of them a rocks glass will serve just as well. If you do go to the extra effort of finding the bottles, why not also pick up some novelty straws to complete the festive look of this serve?

BLACK OLIVE APERITIVO

I can't be trusted around olives. In fact, place a bowl near me and they'll be gone before anyone else gets a look in. This aperitivo-style drink was created in homage to this beloved ingredient. The idea sprang from a paring of Martini's unique and moreish Bianco Vermouth and olives, sipped and savoured one spring evening. I love the interaction between the various spices, woods and herbs in this particular vermouth and find it quite refreshingly (and dangerously) quaffable. (See also Eden, page 64.)

FOR THE COCKTAIL

50 ml/1¾ fl oz Black Olive Martini Bianco (see page 162)
1–2 bar spoons orgeat (almond syrup)
50 ml/1¾ fl oz tonic water
50 ml/1¾ fl oz prosecco
rosemary sprig and black olive, to garnish

To make the cocktail, place the Black Olive Martini Bianco and orgeat in the base of a large wine glass. Two-thirds fill the glass with good-quality cube ice. Stir to combine and chill the liquids. Whilst still stirring, pour over the tonic water followed by the prosecco. Top the glass up with fresh ice, if necessary, then garnish with a sprig of rosemary and a black olive and serve.

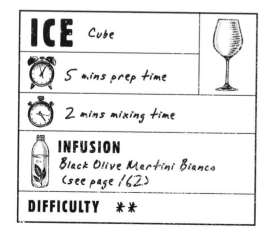

ICE Cube

5 mins prep time

2 mins mixing time

INFUSION
Black Olive Martini Bianco
(see page 162)

DIFFICULTY ✱✱

PISCOLITA

Pisco is such an underused spirit and its fragrance works perfectly in this cocktail. I love the marriage of the light and fresh melon and coconut, bolstered with the earthy, charred notes of black cardamom. See image page 107.

FOR THE COCKTAIL

25 ml/¾ fl oz soda water
35 ml/1¼ fl oz pisco
35 ml/1¼ fl oz galia melon juice
20 ml/¾ fl oz Coconut Syrup (see below)
2 dashes Black Cardamom & Pineapple Bitters
   (see page 29)
2 pineapple leaves, to garnish

FOR THE COCONUT SYRUP

150 ml/5 fl oz coconut water
150 g/5½ oz caster (superfine) sugar

To make the Coconut Syrup, pour the coconut water into a large bowl and add the sugar. Stir until the sugar has completely dissolved, then decant into a bottle and reserve in the fridge until required.

To make the cocktail, fill a highball glass with good-quality cube ice, then add the soda water. Add the remaining ingredients and stir with a bar spoon to combine. Top up the glass with fresh ice, if necessary, then garnish with the pineapple leaves and serve.

VODKA–BAY SALTY DOG

Transforming a simple but classic serve is a joy, especially when the end result is even better than the original as in the case of this drink. The simple vodka and grapefruit drink with a salt rim – AKA the Salty Dog – is given added dimension by infusing the aromatic flavours of vanilla, bay and birch into the vodka before mixing.

FOR THE COCKTAIL

35 ml/1¼ fl oz Bay, Birch & Vanilla Vodka (see page 162)
25 ml/¾ fl oz sugar syrup (see page 26)
50 ml/2 fl oz grapefruit juice
4 drops Vanilla Bitters (see page 28)
lemon wedges and sea salt flakes, for the glass
grilled lemon wheel and bay leaf, to garnish

To prepare your glass, rub a cut lemon wedge around the rim, then dip into a bowl of sea salt flakes. Fill your glass with cube ice and set aside while you mix the cocktail.

To make the cocktail, place all the ingredients in a shaker or stirring jug and fill with good-quality cube ice. Stir until icy cold, then strain into your prepared glass over fresh cube ice. Add more ice to the glass, if necessary, garnish with a grilled lemon wheel and bay leaf, then serve.

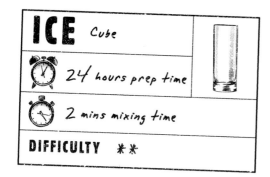

ICE Cube
24 hours prep time
2 mins mixing time
DIFFICULTY ✱✱

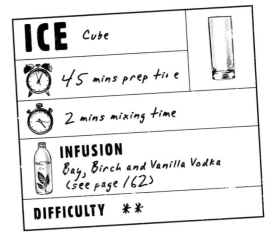

ICE Cube
45 mins prep time
2 mins mixing time
INFUSION
Bay, Birch and Vanilla Vodka
(see page 162)
DIFFICULTY ✱✱

AMERICANO AMERICANO

I love the autumnal colour of this drink and although it conjures up memories of the colder months with its dark red hues, it is an all-year-round sundown sipper. Based, of course, on the classic Americano cocktail, this recipe adds a home-made coffee liqueur to the traditional Campari and sweet vermouth highball.

FOR THE COCKTAIL

25 ml/¾ fl oz Campari
25 ml/¾ fl oz sweet vermouth
60 ml/2 fl oz Coffee Liqueur (see page 160)
50 ml/2 fl oz soda water
orange wedge, to garnish

Place the Campari, sweet vermouth and Coffee Liqueur in a highball or Collins glass and fill with good-quality cube ice. Top the glass up with soda and stir with a bar spoon until well combined and icy cold. Garnish with an orange wedge and serve.

Why not try the coffee liqueur in an espresso Martini? Add 35 ml/1¼ fl oz of the liqueur to 35 ml/1¼ fl oz vodka. Add a double espresso and 15 ml/½ fl oz of sugar syrup. Shake hard over ice and double strain.

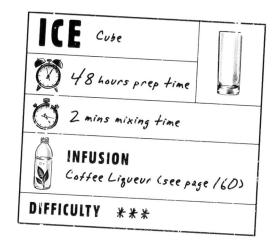

ICE Cube

48 hours prep time

2 mins mixing time

INFUSION
Coffee Liqueur (see page 160)

DIFFICULTY ✳✳✳

Image (from left to right):
Double Black Velvet
(page 114) and
Piscolita (page 104)

BIODYNAMIC SOUR

The use of the yogurt in this 'sour' spin adds additional dimensions of texture by way of depth and effervescence. I love how the sourness is subdued by the cacao liqueur but then lifted further by the aromatic ground cacao nibs that sit atop the drink.

FOR THE COCKTAIL

35 ml/1¼ fl oz Caraway Tequila (see page 162)
15 ml/½ fl oz Briottet white crème de cacao
25 ml/¾ fl oz sugar syrup (see page 26)
15 ml/½ fl oz lemon juice
2 bar spoons Yogurt Syrup (see page 95)
ground cacao nibs, to garnish

To make the cocktail, place all the ingredients in a shaker, then fill with good-quality cube ice and shake hard until icy cold. Double strain the drink into your rocks glass over a block of ice and garnish with freshly ground cacao nibs. Serve.

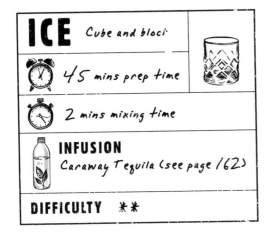

ICE Cube and block

45 mins prep time

2 mins mixing time

INFUSION Caraway Tequila (see page 162)

DIFFICULTY ✳✳

CARAMELIZED RED ONION MANHATTAN

At first glance, you'd be forgiven for thinking that I've lost my mind – 'Onion! In a Manhattan? He's messed up a Gibson, surely'. But despite the juxtaposed nature of this drink, this was no accident – happy or otherwise. The idea behind this was to create flavours similar to those enjoyed whilst eating a roast dinner. The sweetness of the caramelized onions, despite their sharp nature, adds a layer of complexity to this drink that melds with the flavour of sage to create something deliciously savoury and redolent of delicious sage and onion stuffing. Don't be fooled though, if you're too heavy handed with the onion it will overpower the drink – balance is key!

FOR THE COCKTAIL

60 ml/2 fl oz Bourbon
25 ml/¾ fl oz sweet vermouth
2 bar spoons Caramelized Red Onion Reduction
  (see page 163)
skewered pickle onion and sage leaf, to garnish

To make the cocktail, place all the ingredients in a shaker or mixing tin and fill with good-quality cube ice. Stir with a bar spoon until icy cold, then gently strain into a pre-chilled Martini glass. Garnish with a skewered pickled onion and sage leaf and serve.

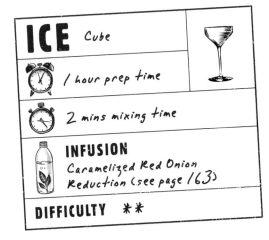

ICE Cube

1 hour prep time

2 mins mixing time

INFUSION
Caramelized Red Onion
Reduction (see page 163)

DIFFICULTY ✳✳

YUZU MARTINEZ

I created this drink for the SUSHISAMBA locations in the USA. Yuzu is a Japanese citrus fruit that I absolutely love. It is more acidic than a lemon or lime and has a unique mandarin-like aroma. The smallest amount of its juice can transform a drink and dropping its zest into a bottle of gin or vodka and leaving it to infuse yields really special results. Though whole fruits are hard to come by, its juice is more readily available (make sure to read the small print as some varieties have been cut with other citrus and are less punchy as a result).

FOR THE COCKTAIL
35 ml/1¼ fl oz Yuzu Gin (see page 163)
25 ml/¾ fl oz Martini Rosso
15 ml/½ fl oz Yuzu Liqueur (see page 163)
3 bar spoons maraschino liqueur

To make the cocktail, place all the ingredients in a shaker or mixing tin and fill with good-quality cube ice. Stir with a bar spoon until icy cold, then gently strain into a pre-chilled Martini glass. Serve.

This drink uses the yuzu zest, which you can pick up from Asian supermarkets or well-stocked delis.

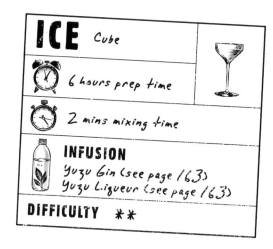

ICE Cube

6 hours prep time

2 mins mixing time

INFUSION
Yuzu Gin (see page 163)
Yuzu Liqueur (see page 163)

DIFFICULTY ★★

DOUBLE BLACK VELVET

The fruity sweetness of the blackberry perfectly complements and lifts the earthy notes of the stout in this cocktail. Though Guinness is notoriously heavy, when mixed with fizz it is transformed into a bubbly and effervescent brunch-time serve. See image page 106.

FOR THE COCKTAIL

50 ml/1¾ fl oz Blackberry & Guinness Reduction
   (see below)
100 ml/3½ fl oz Champagne

FOR THE BLACKBERRY & GUINNESS REDUCTION

440 ml/14 fl oz Guinness
70 g/2½ oz caster (superfine) sugar
30 g/1 oz blackberries

To make the Blackberry & Guinness Reduction, place the Guinness in a pan over a medium to low heat and cook until it has reduced by half (do not attempt to rush this process over a high heat as the ale will lose flavour.) Once reduced, add the sugar and stir until dissolved, then remove from the heat and stir in the blackberries. Set aside to infuse and cool to room temperature. Once cooled, filter the liquid through a fine muslin-lined sieve (strainer) and reserve the liquid. Transfer to the fridge and reserve until required.

To make the cocktail, place the Blackberry & Guinness Reduction in the base of a Champagne flute and top up with the Champagne. Gently stir to combine and serve.

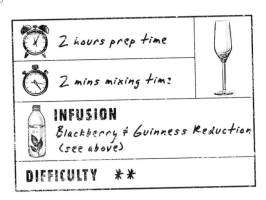

2 hours prep time

2 mins mixing time

INFUSION
Blackberry & Guinness Reduction (see above)

DIFFICULTY ✱✱

WINTER-SPICED CHAMPAGNE COCKTAIL

Nothing heralds the start of the festive season better than the arrival of rich and fragrant mince pies on the supermarket shelves. Biting through the buttery pastry and first tasting that boozy, gooey filling immediately sends me giddy with childlike excitement at the prospect of Santa's imminent arrival.

This cocktail is essentially a festive take on the classic Champagne cocktail. It is great to serve to family or friends at a Christmas gathering and is both light and warming, making it perfect for both daytime and evening occasions.

FOR THE COCKTAIL
65 ml/2¼ fl oz Winter-spiced Liqueur (see page 164)
65 ml/2¼ fl oz Champagne

To make the cocktail, place the Winter-spiced Liqueur in a shaker or mixing tin, then add a little good-quality cube ice (2 or 3 cubes will suffice). Stir with a bar spoon until icy cold, then strain into a Champagne flute. Top the glass up with Champagne and gently stir to combine. Serve.

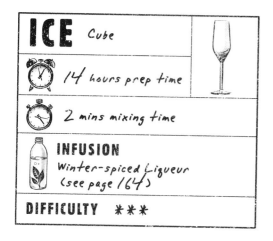

ICE Cube

14 hours prep time

2 mins mixing time

INFUSION Winter-spiced Liqueur (see page 164)

DIFFICULTY ✳✳✳

Though the liqueur in this drink has been created for this particular cocktail, it can also be chilled and sipped neat, added to a shot of gin, poured over ice or sneaked into a Manhattan for a festive spin on another classic. However you use it, it's great for those colder months.

PINK PEPPERCORN LEMONADE

This is a fresh and refreshing serve for a warm day or balmy evening. The fresh zestiness of the grapefruit lifts the warming blend of pink peppercorn and elderflower. Though the recipe below is for a single serve, this can easily be batched with the addition of a little soda in each glass to finish. .

FOR THE COCKTAIL

35 ml/1¼ fl oz Pink Peppercorn Gin (see page 164)
25 ml/¾ fl oz Clarified Grapefruit Juice (see page 36)
15 ml/½ fl oz sugar syrup (see page 26)
15 ml/½ fl oz elderflower cordial
1 dash lemon bitters
35 ml/1¼ fl oz soda water
long length of lemon peel, rolled and secured with a tooth pick
     to resemble a rose, to garnish

To make the cocktail, place the Pink Peppercorn Gin, Clarified Grapefruit Juice, sugar syrup, elderflower cordial and lemon bitters in the base of a highball glass and half-fill with good-quality cube ice. Stir with a bar spoon to meld the flavours, then, while still stirring, add the soda water. Top up the glass with more ice and garnish with a lemon rose. Serve.

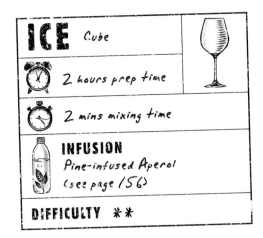

ICE Cube

2 hours prep time

2 mins mixing time

INFUSION
Pine-infused Aperol
(see page 156)

DIFFICULTY **

To make a lemon rose, carefully peel a lemon from top to bottom in one long strip of rind, then roll to create a spiral rose shape and secure with a cocktail stick. These also make great visuals when making your own ice cubes (see page 30).

ACORN APERITIVO

This is one of those drinks where the first sip has always stayed with me: the initial hit of mouth-puckering grapefruit oil leading through to a creamy sweetness reminiscent of freshly baked bread and finishing with the light bitter-sweet undertones of the Cocchi Rosa. Straightaway, I knew I had stumbled upon something special, and I have loved it ever since.

FOR THE COCKTAIL

50 ml/1¾ fl oz Grey Goose vodka
30 ml/1 fl oz Acorn Liqueur (see page 164)
15ml/½ fl oz Cocchi Rosa
coin of grapefruit zest, to garnish

Place the vodka, Acorn Liqueur and Cocchi Rosa in a shaker or mixing jug and top up with good-quality cube ice. Stir the mixture with a bar spoon until icy cold, then strain into a small Martini glass and serve garnished with a coin of grapefruit zest.

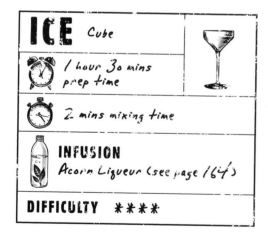

ICE Cube

1 hour 30 mins prep time

2 mins mixing time

INFUSION
Acorn Liqueur (see page 164)

DIFFICULTY ✳✳✳✳

KOBE COCKTAIL

In 2009, a very well-known New York bar called Please Don't Tell (PDT) put a cocktail called the Benton's Old Fashioned on their menu. At the time, this drink would have caused a little bit of fuss as it contained a fat-washed bacon infusion. I first put this delightful riff on the SUSHISAMBA London menu back in 2014, then at the New York West Village location in early 2015 (though this was with Wagyu fat). The additional marbling of this grade of beef gives exquisite complexity and mouthfeel. (Fat washing is also used in the Roasted Cosmo page 94 , Chocolate & Blue Cheese Martini page 128 and Bulletproof Old Fashioned page 133.) Though this drink is totally unique in its flavour, it owes it all to the Benton's Old Fashioned from PDT.

FOR THE COCKTAIL:

60 ml/2 fl oz Kobe/Wagyu Whisky (see page 165)
2–3 bar spoons maple syrup, depending on
    sweetness desired
5 drops Burnt Caramel Bitters (see page 28)

To make the cocktail, place all the ingredients in a mixing tin or glass and fill with good-quality cube ice. Using a bar spoon, stir until icy cold, then gently strain into your desired glass over a large ice block. Serve.

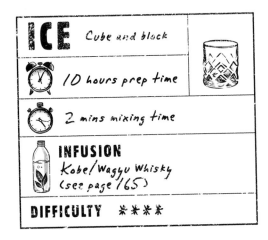

ICE Cube and block

10 hours prep time

2 mins mixing time

INFUSION Kobe/Wagyu Whisky (see page 165)

DIFFICULTY ★★★★

Fat washing is everywhere and the principle is fairly straightforward. Take an ingredient high in fat, e.g. bacon, cheese, butter or bone marrow, render it if needed, then simply infuse and filter the liquid. Nowadays, some variations may require the infusion to be cooked lightly. Others may use the sous vide technique. Most if not all the fat washing I have done ends with freezing the product, then filtering it to retain a velvety but clear-looking liquid.

ESPRESSO GIN & TONIC

The ultimate hipster cocktail. You won't find two trendier drinks than gin and coffee. This is great as a single serve, but also a brilliantly simple to scale up if you're making drinks for a large group of friends (everyone knows how to make a G&T, right?!) This can be mixed together before drinking to meld the flavours or sipped through a straw so that the coffee hit is saved right until the end.

FOR THE COCKTAIL

50 ml/1¾ fl oz Pine Needle Gin (see page 165)
1 bar/teaspoon sugar syrup (see page 26)
tonic water, to top up
single shot of fresh espresso (30 ml/1 fl oz), cooled to room temperature
pine sprig, to garnish

To make the cocktail, place the Pine Needle Gin and sugar syrup in the base of a highball glass and two-thirds fill with good-quality cube ice. Top the glass up with tonic water to three-quarters full, then stir with a bar spoon to combine. Add a small sprig of pine to the glass, then gently tip the espresso over the top of the drink so that it floats on top of the drink and slowly seeps down. Enjoy.

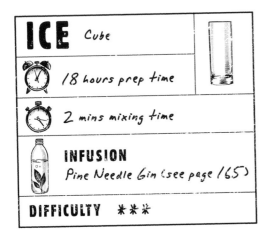

ICE Cube

18 hours prep time

2 mins mixing time

INFUSION
Pine Needle Gin (see page 165)

DIFFICULTY ✳✳✳

Pine needle essence can be used to make the gin in place of fresh pine leaves; simply follow the recipe for Pine-infused Aperol on page 156 and substitute the Aperol for gin.

ICONOCLASTIC

DECEPTION
AKA 'CLARIFIED HEMINGWAY'

I love the deceptiveness of using misdirection when creating drinks and the ability to create a multi-sensory experience. This Hemingway-daiquiri riff is a great example of this as not only does it engage our sense of flavour and taste, but also plays with sight and mouthfeel.

This is one of my all-time favourite cocktails. I love the clarified nature of this drink, which leaves it ever so light in colour, bursting with flavour and deceptively translucent.

FOR THE COCKTAIL

50 ml/1¾ fl oz good-quality light rum, such as
    Bacardi Carta Blanca
15 ml/½ fl oz maraschino liqueur
35 ml/1¼ fl oz Clarified Grapefruit Juice
    (see page 36)
15 ml/½ fl oz Clarified Lime Juice (see below)
1 bar spoon sugar syrup (see page 26)
maraschino cherry, to garnish

FOR THE CLARIFIED LIME JUICE

Yields 750 ml/1¼ pints

1 litre/1¾ pints lime juice
8 g/¼ oz agar agar

To make the Clarified Lime Juice, pour the lime juice into a pan and place over a medium heat. Heat until just simmering, then gradually pour in the agar agar and stir until completely dissolved. Remove the pan from the heat and allow to cool to room temperature, then transfer to a non-reactive container with a lid and place in the freezer, covered, for 24 hours.

The next day, remove the frozen lime juice from the freezer and leave to melt through a muslin-lined sieve (strainer) or coffee filter into a large bowl. Decant the liquid into a bottle and reserve in the fridge until required.

To make the cocktail, place all the ingredients in a mixing glass or tin, then two-thirds fill with good-quality cube ice. Stir the drink with a bar spoon until icy cold, then strain into a small coupette or Martini glass. Garnish with a maraschino cherry and serve.

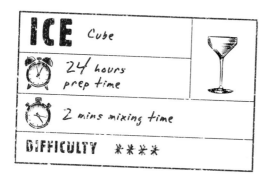

ICE *Cube*

24 hours prep time

2 mins mixing time

DIFFICULTY ✱✱✱✱

CHOCOLATE & BLUE CHEESE MARTINI

One night in 2014, having arrived home after a long and late shift, I opened the fridge to find it almost bare, except for a packet of chocolate digestive biscuits and some blue cheese (never send a hungry man food shopping). Part out of curiosity, but mostly out of hunger, I spread some cheese over the digestives and that was that – an idea was born. Later that year I entered the result of that experiment into a global cocktail competition and won my national final.

The process by which flavour is added to this cocktail is called fat washing. In much the same way as you remove fat from gravy, you separate the two elements. The result is a rich, flavoursome and clear liquid. (Fat washing is also used in the Roasted Cosmo page 94, Kobe Cocktail page 120 and Bulletproof Old Fashioned page 133.)

FOR THE COCKTAIL

50 ml/1¾ fl oz Blue Cheese Gin (see page 165)
2 bar spoons of Mozart Chocolate Vodka
1 bar spoon Briottet dark crème de cacao
1 bar spoon Briottet white crème de cacao
1 bar spoon sugar syrup (see page 26)
2 bar spoons dry vermouth
4 drops olive oil, to garnish
2 drops chocolate bitters, to garnish

To make the cocktail, place all the ingredients in a mixing glass or tin, then two-thirds fill with good-quality cube ice. Stir the drink with a bar spoon until icy cold, then gently strain into a small coupette or Martini glass. Using a pipette, place the droplets of olive oil on the top of the drink and then a smaller droplet of chocolate bitters in the centre of the olive oil. Serve.

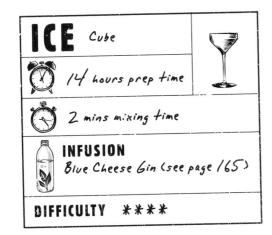

ICE Cube

14 hours prep time

2 mins mixing time

INFUSION
Blue Cheese Gin (see page 165)

DIFFICULTY ✱✱✱✱

FILTHY MARTINI

Following a trend for more savoury cocktails, this is a marriage of the Gibson and Dirty Martini. Using two of my favourite condiments I created a culinary wholegrain mustard and Branston Pickle martini. A real Marmite drink, but for those (like me!) who love mustard, pickle and oysters, this is heaven in a glass.

FOR THE COCKTAIL

35 ml/1¼ fl oz gin
25 ml/¾ fl oz Wholegrain Mustard Vodka (see page 166)
15 ml/½ fl oz Branston Pickle Vodka (see page 166)
15 ml/½ fl oz dry vermouth
1 oyster, to garnish (optional)

To make the cocktail, place all the ingredients in a mixing glass or tin, then two-thirds fill with good-quality cube ice. Stir the drink with a bar spoon until icy cold, then gently strain into a coupette or Martini glass.

ICE Cube

2 hours prep time

2 mins mixing time

DISTILLATION
Wholegrain Mustard Vodka
(see page 166)
Branston Pickle Vodka
(see page 166)

DIFFICULTY ✱✱✱✱✱

For an extra savoury kick, place a freshly shucked oyster in the base of the glass before pouring over the cocktail.

BULLETPROOF OLD FASHIONED

For those who have not tried Bulletproof Coffee, it really is something else – a richer, more intense, single-origin coffee. Free from mycotoxins, it is brewed and then blended with unsalted grass-fed butter. The result is the creamiest and most delicious cup of coffee, you've ever tasted.

Developing this idea into a cocktail was a whole lot of fun. Deciding to use the butter as the principle flavour in this drink, rather than the coffee, added an even creamier finish, which when blended with Tequila produced an epic result. Take the time to look for the right grass-fed butter and avoid salted varieties as it does make the difference.

FOR THE COCKTAIL

60 ml/2 fl oz Butter-washed Tequila (see page 162)
2 bar spoons maple syrup
10 ml/0.35 fl oz Ristretto Vodka (see page 166)
5–8 drops Burnt Caramel Bitters (see page 28)

To make the cocktail, place all the ingredients in a mixing glass or tin, then fill with good-quality cube ice. Stir the drink with a bar spoon until icy cold, then gently strain into a rocks glass filled with good-quality ice cubes or block ice.

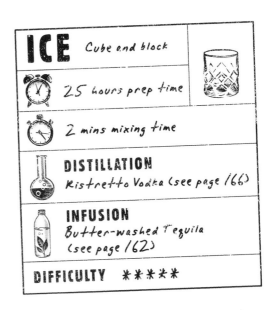

ICE Cube and block

2.5 hours prep time

2 mins mixing time

DISTILLATION
Ristretto Vodka (see page 166)

INFUSION
Butter-washed Tequila (see page 162)

DIFFICULTY ✱✱✱✱✱

PARSLEY

The parsley in this cocktail brings out the greener vegetable notes of the banana liqueur. There is a fine balance of flavours here, so be extra careful with your measurements as too much of the parsley or banana can easily overpower the other. The lemon zest garnish may seem inconsequential, but is in fact vital as it both balances the sweetness of the liqueur and subdues accents of the parsley.

FOR THE COCKTAIL

60 ml/2 fl oz Parsley Vodka (see page 167)
20 ml/¾ fl oz Briottet crème de banane
15 ml/½ fl oz Cocchi Americano
coin of lemon zest, to garnish

To make the cocktail, place all the ingredients in a mixing glass or tin, then two-thirds fill with good-quality cube ice. Stir the drink with a bar spoon until icy cold, then strain into a small Martini glass. Squeeze a coin of lemon zest to spray the juices over the drink and stem of the glass, then place on the surface of the drink and serve.

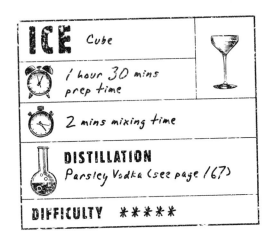

ICE Cube

1 hour 30 mins prep time

2 mins mixing time

DISTILLATION
Parsley Vodka (see page 167)

DIFFICULTY ★★★★★

PEANUT – RYE & DRY

Based, albeit loosely, on a classic Godfather, this drink was created to emulate the richness of peanut butter. The addition of vermouth not only adds a dry note, but its nut-like characteristics complement the peanut butter liqueur.

FOR THE COCKTAIL

35 ml/1¼ fl oz Peanut Butter Liqueur (see page 167)
25 ml/¾ fl oz rye whiskey
1 bar spoon dry vermouth

To make the cocktail, place all the ingredients in a mixing glass or tin, then fill with good-quality cube ice. Stir the drink with a bar spoon until icy cold, then strain gently into a rocks glass filled with good-quality ice cubes or chipped ice.

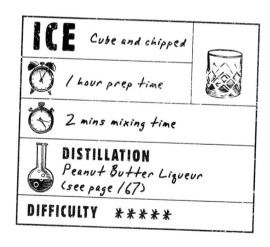

ICE Cube and chipped

1 hour prep time

2 mins mixing time

DISTILLATION
Peanut Butter Liqueur
(see page 167)

DIFFICULTY ✱✱✱✱✱

EPIPHANY

A moment of realization. Despite its light and fresh appearance, this delightful drink has hidden depths. The honey-like aroma of pollen is spritzed with Champagne and enriched with base notes of caramel and maple.

FOR THE COCKTAIL
35 ml/1¼ fl oz Pollen Vodka (see page 167)
10 drops Burnt Caramel Bitters (see page 28)
2–3 bar spoons Honey Maple Syrup (see below)
90 ml/3 fl oz Champagne

FOR THE HONEY MAPLE SYRUP
Yields 150 ml/5½ fl oz

100 ml/3½ fl oz maple syrup
50 ml/1¾ fl oz runny honey

To make the Honey Maple Syrup, combine the maple syrup and honey in a bowl and mix until well combined. Set aside until required, mixing again just before using.

To make the cocktail, place the Pollen Vodka, Burnt Caramel Bitters and Honey Maple Syrup in a Champagne flute, then top up with the Champagne. Gently stir to combine, then serve.
.

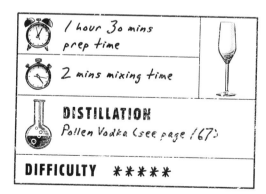

1 hour 30 mins prep time

2 mins mixing time

DISTILLATION
Pollen Vodka (see page 167)

DIFFICULTY ✳✳✳✳✳

SESAME

The mellow sweetness of coconut makes a brilliant partner to the earthy richness of Scotch in this quenching drink. Toasted sesame oil gives an additional layer of underlying nuttiness that rounds off the drink to create something that is sweet, savoury and refreshing all at once.

FOR THE COCKTAIL

60 ml/2 fl oz Sesame Scotch Distillate (see page 168)
1 bar spoon Talisker Whisky
90 ml/3 fl oz coconut water
30 ml/1 fl oz Coconut Sugar Syrup (see below)
coconut shavings and mint sprig, to garnish

FOR THE COCONUT SUGAR SYRUP

Yields 250 ml/9 fl oz

150 ml/5 fl oz coconut water
150 g/5½ oz caster (superfine) sugar

To make the Coconut Sugar Syrup, combine the ingredients in a large jug or bowl and stir until the sugar has dissolved. Decant into a bottle and reserve in the fridge until required.

To make the cocktail, two-thirds fill a highball glass with good-quality cube or chipped ice and add the ingredients. Stir with a bar spoon until icy cold, then serve garnished with coconut shavings and a large mint sprig (agitated to release its aroma).

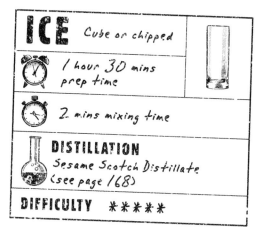

ICE Cube or chipped

1 hour 30 mins prep time

2 mins mixing time

DISTILLATION
Sesame Scotch Distillate
(see page 168)

DIFFICULTY ✱✱✱✱✱

UMAMI MARY

This is my take on everyone's brunch-time favourite. Great with vodka and gin, it is also amazing with Tequila. This version uses the nutty and savoury notes of Shiitake mushrooms distilled into vodka which, when added to a spice mix blend and topped with quality tomato juice, makes for one hell of a Bloody Mary.

FOR THE COCKTAIL

35 ml/1¼ fl oz Shiitake Vodka (see page 168)
2 bar spoons lemon juice
20 ml/¾ fl oz Spice Mix (see below)
100 ml/3½ fl oz tomato juice
caper berries and lemon wedges, to garnish

FOR THE SPICE MIX

Yields 450 ml/15 fl oz

60 g/2 oz black peppercorns
30 g/1 oz green peppercorns
30 g/1 oz dried lemon peel
20 g/¾ oz dried ancho chillies
4 g/¼ oz cumin seeds
10 g/ ¼ oz sea salt flakes
185 g/7 oz red chillies, sliced
100 g/3½ oz jalapeño chillies, sliced
70 g/2½ fl oz Red Tabasco Sauce
60 ml/2 fl oz Green Tabasco Sauce
260 ml/8¾ fl oz soy sauce
75 ml/2½ fl oz malt vinegar
100 ml/3½ fl oz jalapeño pickle brine
125 ml/4 fl oz water

To make the Spice Mix, place the peppercorns, lemon peel, dried chillies, cumin seeds and sea salt in a blender and pulse until well combined. Add the red and jalapeño chillies and pulse again to combine and break down the chillies. Add the remaining ingredients and blend again until well combined, then transfer to a non-reactive container with a lid and place in the fridge, covered, for at least 24 hours.

The next day, pass the mixture through a fine sieve (strainer) into a large bowl and reserve in the fridge until required.

To make the cocktail, two-thirds fill a highball glass with good-quality cube ice and add the ingredients. Stir with a bar spoon until icy cold, then serve garnished with a caper berry and lemon wedge.

If you prefer your Marys slightly spicier, simply add more of the spice mix – 30 ml/1 fl oz should do the job.

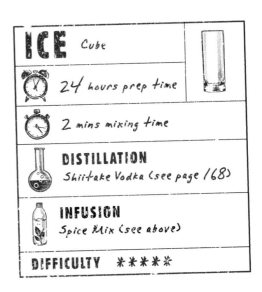

ICE	Cube	
	24 hours prep time	
	2 mins mixing time	
DISTILLATION	Shiitake Vodka (see page 168)	
INFUSION	Spice Mix (see above)	
DIFFICULTY	★★★★☆	

GREEN APPLE RYE

Cucumbers, apples and elderflowers are all amazingly light, fresh and crisp on their own and even better together, Here, these summery flavours are rounded out with a deliciously mellow apple-infused rye that adds a real depth of flavour to this rich yet bright and refreshing drink.

FOR THE COCKTAIL

45 ml/1½ fl oz Green Apple Rye (see page 168)
15 ml/½ fl oz St. Germain Elderflower Liqueur
15 ml/½ fl oz sugar syrup (see page 26)
2 dashes Cucumber Bitters (see page 29)
1 dash Malic Acid Solution (see below)
60 ml/2 fl oz prosecco
long cucumber peel, to garnish

FOR THE MALIC ACID SOLUTION

50 ml/5 fl oz vodka
35 g/1¼ oz malic acid powder

To make the Malic Acid Solution, combine the vodka and malic acid in a jug and stir until the acid has completely dissolved. Decant into a bottle and reserve in the fridge until required.

To make the cocktail, wrap the cucumber peel garnish around the inside of a Collins glass and fill with good-quality cube or chipped ice. Place the remaining ingredients except the prosecco, in a mixing glass or tin and fill with good-quality cube ice, then stir with a bar spoon until icy cold. Add the prosecco, then gently roll the tin to combine. Strain into the serving glass, top up with more fresh ice, if necessary, and serve.

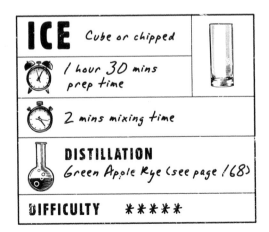

ICE Cube or chipped

1 hour 30 mins prep time

2 mins mixing time

DISTILLATION
Green Apple Rye (see page 168)

DIFFICULTY ★★★★★

DEFECTED BLACK RUSSIAN

This drink was created for a recent menu I conceived at Duck & Waffle entitled 'Colour Blind'. Essentially it was designed to show how colour affects our perception of flavour and how, much like the impact of window dressing, we buy with our eyes.

FOR THE COCKTAIL

35ml Ristretto Vodka (see page 166)
20ml Black Cow Vodka
15 ml/½ fl oz Removed Kahlua (see page 169)
15ml Briottet white crème de cacao
1 bar spoon sugar syrup (see page 26)

FOR THE COLA PAINT

150 ml/5 fl oz cola (ideally flat)
2 g/0.07 oz pectin powder
(approximately 1.5–3 % of the cola weight)

To make the Cola Paint, place the cola in a microwave-proof container and pulse with a stick (immersion) blender to remove as much carbonation as possible. Weigh the liquid and add 1 % of pectin to the total volume. Transfer to the microwave and cook on medium to high for 1 minute. Pulse the liquid again to ensure the pectin has dissolved. The liquid should now appear thicker – if not, repeat the process, adding another 0.5 % of pectin. Strain through a fine sieve (strainer) and transfer to the fridge for at least 3 hours to thicken further.

An hour before you want to make your cocktail, place your glasses in the freezer. Once frozen, paint a stripe of Cola Paint down the inside of each. Return the glasses to the freezer while you make the cocktail. To make the cocktail, place all the ingredients, except the paint, in a mixing glass or tin and fill with good-quality cube ice, then stir with a bar spoon until icy cold. Gently strain into a small Martini glass and serve.

HAY

Here, Jack Daniels is infused with the mellow earthiness of hay and lifted with the bitter-sweetness of caramel. One of my favourites, it's a great late-night drink for when you are putting the world to rights with friends.

FOR THE COCKTAIL

50 ml/1¾ fl oz Hay Jack Daniels (see page 169)
1 bar spoon maple syrup
8–1- drops Burnt Caramel Bitters (see page 28)

To make the cocktail, place all the ingredients in a mixing glass and fill with good-quality cube ice, then stir with a bar spoon until icy cold. Place a large block of ice in a rocks glass, then gently strain the cocktail over and serve.

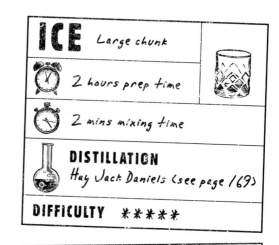

ICE Large chunk
2 hours prep time
2 mins mixing time
DISTILLATION
Hay Jack Daniels (see page 169)
DIFFICULTY ★★★★★

ICE Cube
2 hours prep time
2 mins mixing time
DISTILLATION
Ristretto Vodka (see page 166)
Removed Kahlua (see page 169)
DIFFICULTY ★★★★★

Image: Hay

WASABI COCKTAIL

This drink is based on a vesper, the cocktail made famous by 007 in Ian Fleming's *Casino Royale*. In my version, the fiery and earthy flavour of wasabi is calmed and enhanced by the cucumber notes in the Hendrick's Gin for a subtle, well-balanced finish that titillates the palate without overpowering.

FOR THE COCKTAIL

30 ml/1 fl oz Hendrick's Gin
30 ml/1 fl oz Wasabi Vodka (see page 169)
30ml/1 fl oz Lillet Blanc Vermouth
4 drops Cucumber Bitters (see page 29)
long cucumber peel, to garnish

To make the cocktail, place all the ingredients in a mixing glass or tin, then fill with good-quality cube ice. Stir the drink with a bar spoon until icy cold, then double strain into a small Martini glass. Serve garnished with a spiral of cucumber peel twisted around the inside of the glass.

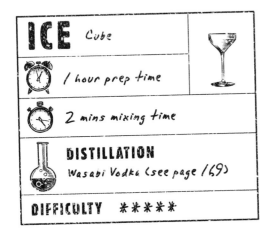

ICE Cube

1 hour prep time

2 mins mixing time

DISTILLATION
Wasabi Vodka (see page 169)

DIFFICULTY *★★★★★*

Why not try using the Wasabi Vodka in the Clarified Bloody Mary (page 72), for an alternative sipping Mary-style cocktail.

DISTILLATIONS AND INFUSIONS

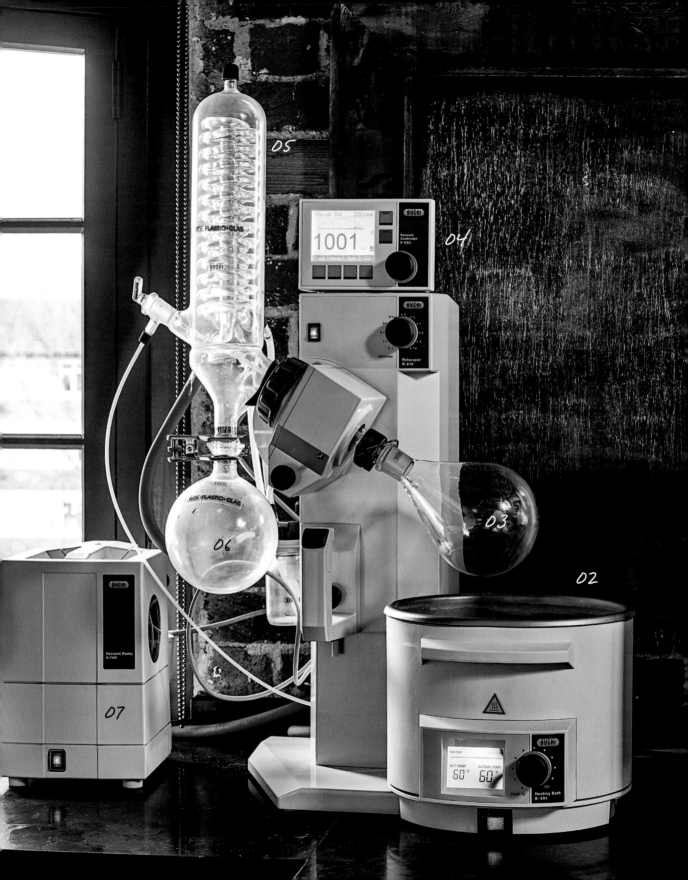

01
Circulating Chiller
Contains water that is cooled and recirculated through the condensing coil.

02
Water Bath
Contains an element which heats water to a precise temperature, allowing for a more controlled circulation surrounding the evaporation flask.

03
Evaporating Flask
Contains the infusion you wish to distil, be it water or alcohol-based and containing the compounds for distillation.

04
Rotary Evaporator Controller
Controls and regulates the pressure at which distillations are evaporated, as well as enabling you to pre-set distillation times for precise repetitive evaporation.

05
Condensing Coil
As the evaporated infusion hits the chilled condensing coil, it transforms back into a liquid state and collects in the receiving flask.

06
Receiving Flask
Collects the now distilled infusion.

07
Vacuum Pump
A diaphragm pump which controls the speed and pressure at which the distillation is extracted by way of vacuum.

DISTILLATION AND VACUUM EVAPORATION

The Rotary Evaporator is one of the most elaborate pieces of kit that I use when making drinks. The equipment is very expensive, so possibly not entirely suitable for the home bartender, but the results it produces are insanely good. It enables you to distil alcohol, or any other liquid, with various compounds or ingredients to create levels of flavour that simply wouldn't be possible otherwise. The process involves reducing the pressure of a liquid and the components within it in order to lower its boiling point. The liquid then evaporates to form a gas that is in turn cooled to return it to a liquid state, much like the traditional boiling method. This end liquid, or distillate, has now taken on the intense flavour of the original infusion and compounds placed within it. The process is explained in steps below.

- Alcohol and the compounds you want to flavour it with (this can vary from herbs and spices to manmade products such as peanut butter) are placed in the evaporation flask.
- The water bath is preheated to a precise temperature before the evaporation flask is lowered into it.
- The vacuum pump applies vacuum through the system to lower the pressure of the distillation, which in turn lowers its boiling point.
- The rotary evaporator lowers the evaporation flask into the heated water bath and spins the compounds within to ensure they are evenly heated.
- As the liquid in the evaporation flask comes to the boil it starts to evaporate and the gas is taken up through the condenser where it is then chilled with liquid from the recirculating chiller.
- As the gas is cooled it returns to a liquid state and drips down through the condenser into the receiving flask.
- The distillate is now highly concentrated with the flavour of the original compounds. The strength of the distillate can now be rectified with mineral water and it is then ready for use.

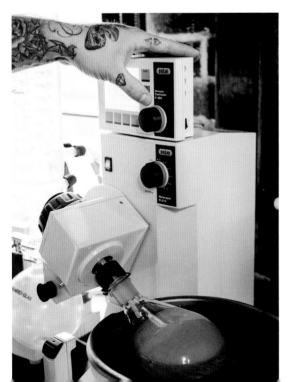

INFUSIONS

CORIANDER GIN

Yields approximately 500 ml/18 fl oz

500 ml/18 fl oz gin
1 bunch micro coriander (cilantro)

- Place the ingredients in a non-reactive container with a lid and leave to infuse, covered, for at least 1 hour (longer, if you prefer a stronger flavour).
- Once infused, strain through a muslin-lined sieve (strainer) or coffee filter and collect the liquid.
- Decant the liquid into a bottle and reserve in the fridge until required.

BITTER ORANGE NEROLI GIN

Yields approximately 450 ml/16 fl oz

500 ml/18 fl oz gin
10 g/¼ oz dried bitter orange peel

- Place the ingredients in a non-reactive container with a lid and leave to infuse, covered, for at least 5 hours.
- Once infused, strain through a muslin-lined sieve (strainer) or coffee filter and collect the liquid.
- Decant the liquid into a bottle and reserve in the fridge until required.

PINE-INFUSED APEROL

Yields approximately 500 ml/18 fl oz

500 ml/18 fl oz Aperol
8 drops food-grade pine needle essence

- Preheat the water bath to 45°C/113°F.
- Place the ingredients in a vacuum bag and seal.
- Place the vacuum bag in the water bath and leave to cook for 50 minutes.
- Remove the vacuum bag from the water bath and set aside to cool.
- Strain the infusion through a coffee filter into a large jug or bowl.
- Decant the liquid into a bottle and reserve in the fridge until required.

DILL POLLEN-INFUSED SHERRY

Yields approximately 350 ml/12 fl oz

5 g/¼ oz dill pollen
350 ml/12 fl oz fino sherry

- Preheat the water bath to 45°C/113°F.
- Place the ingredients in a vacuum bag and seal.
- Place the vacuum bag in the water bath and leave to cook for 40 minutes.
- Remove the vacuum bag from the water bath and set aside to cool.
- Strain the infusion through a coffee filter into a large jug or bowl.
- Decant the liquid into a bottle and reserve in the fridge until required.

BEETROOT & CHOCOLATE LIQUEUR
Yields 500 ml/18 fl oz

25 g/¾ oz food-grade freeze-dried beetroot (beet)
350 ml/12 fl oz vodka
125 ml/4 fl oz Briottet white crème de cacao
150 ml/5 fl oz sugar syrup (see page 26), or more to taste

- Place the beetroot (beet), vodka and cacao liqueur in a non-reactive plastic container with a lid and leave to infuse, covered, for a minimum of 12 hours, but preferably overnight.
- Once infused, strain the mixture through a tea strainer or fine sieve (strainer) and collect the liquid.
- Add the sugar syrup and stir to combine. If you prefer a sweeter liqueur, add more syrup until you are happy with the taste.
- Decant the liqueur into a bottle and reserve in the fridge until required.

EXPRESSO VODKA
Yields approximately 500 ml/18 fl oz

500 ml/18 fl oz vodka
50 g/1¾ oz walnut shells
100 g/3½ oz dandelion leaves
10g/¼ oz birch bark

- Place the ingredients in a non-reactive container with a lid and leave to infuse, covered, for at least 12 hours.
- Once infused, strain through a muslin-lined sieve (strainer) or coffee filter and collect the liquid.
- Decant the liquid into a bottle and reserve in the fridge until required.

KAFFIR LIME LEAF GIN
Yields 500 ml/18 fl oz

28 g/1 oz kaffir lime leaves, torn
500 ml/18 fl oz gin

- Place the lime leaves and gin in the base of a whipped cream dispenser.
- Seal the dispenser with the lid and charge with one N_2O cartridge.
- Shake hard, then repeat the process with another cartridge.
- Invert the canister and empty the gin into a tall non-reactive container.
- Reserve the infused gin until required.

PEA-INFUSED GIN
Yields approximately 700 ml/1¼ pints

700 ml/1¼ pints gin
225 g/8 oz sugar snap peas, shells snapped to release flavour
10 g/¼ oz mint leaves

- Pour the gin into the non-reactive plastic container with a lid.
- Add the sugar snaps and mint, then secure the lid and place in the fridge for at least 24 hours, rolling or gently aggravating the mix every few hours to infuse.
- Once infused, strain through a muslin-lined sieve (strainer) or coffee filter and collect the liquid.
- Decant the liquid into a bottle and reserve in the fridge for up to 2 weeks, until required.

BLACK CHERRY LIQUEUR

Yields approximately 400 ml/14 fl oz

12 ml/½ fl oz food-grade black cherry essence
350 ml/12 fl oz vodka
100 g/3½ oz caster (superfine) sugar

- Place the cherry essence and vodka in a non-reactive container with a lid and leave to infuse, covered, for 1 hour.
- Once infused, strain through a muslin-lined sieve (strainer) or coffee filter and collect the liquid.
- Stir the sugar into the infused vodka until dissolved, then decant the liquid into a bottle and reserve in the fridge until required.

MARIGOLD VODKA

Yields approximately 500 ml/18 fl oz

500 ml/18 fl oz vodka
10 g/¼ oz marigold petals, washed

- Place the vodka and marigold petals in a non-reactive container with a lid and leave to infuse, covered, for at least 4 hours (longer if you prefer a stronger flavour).
- Once infused, strain through a muslin-lined sieve (strainer) or coffee filter and collect the liquid.
- Decant the liquid into a bottle and reserve in the fridge until required.

COCONUT-WASHED LIGHT RUM

Yields approximately 500 ml/18 fl oz

500 ml/18 fl oz light white rum
2.5 g/0.09 oz food-grade coconut essence

- Place the vodka and coconut essence in an empty bottle or non-reactive container with a lid and seal.
- Roll the container to combine the liquids and reserve in the fridge until required.

HOPPED SCOTCH

Yields approximately 500 ml/18 fl oz

500 ml/18 fl oz Scotch whisky
10 g/¼ oz hops (I like Goldings)

- Preheat the water bath to 50°C/122°F.
- Place the ingredients in a vacuum bag and seal.
- Place the vacuum bag in the water bath and leave to cook for 45 minutes.
- Remove the vacuum bag from the water bath and set aside to cool.
- Strain the infusion through a 100 micron superbag into a large jug or bowl.
- Decant the liquid into a bottle and reserve in the fridge until required.

JALAPEÑO TEQUILA
Yields approximately 500 ml/18 fl oz

500 ml/18 fl oz Reposado Tequila
8–10 pickled jalapeño slices

- Place the Tequila and jalapeño slices in an empty bottle or non-reactive container with a lid and leave to infuse, covered, for at least 12 hours. (If you prefer less heat, infuse for a maximum of 5 hours.)
- Reserve in the fridge until required, straining the jalapeño slices just before using.

PANETTONE-INFUSED COGNAC
Yields approximately 500 ml/18 fl oz

500 g/1 lb 2 oz panettone
700 ml/1¼ pints cognac

- Place the panettone and cognac in a non-reactive container with a lid and leave to infuse, covered, for at least 5 hours or up to overnight.
- Once infused, strain through a sieve (strainer) or coffee filter and collect the liquid, ensuring that you squeeze out and strain any residual liquid held within the pieces of panettone.
- Decant the liquid into a bottle and reserve in the fridge until required.

LIQUEUR DE FLEURS
Yields 500 ml/18 fl oz

500 ml/18 fl oz St. Germain Elderflower Liqueur
15 g/½ oz cornflower heads
10 g/¼ oz marigold petals

- Place the ingredients in a non-reactive container with a lid and leave to infuse, covered, for at least 24 hours.
- Once infused, strain through a muslin-lined sieve (strainer) or coffee filter and collect the liquid.
- Decant the liquid into a bottle and reserve in the fridge until required.

COSMO DE PROVENCE MIX
Yields 500 ml/18 fl oz

350 ml/12 fl oz citron vodka
150 ml/5 fl oz triple sec
10 g/¼ oz dried herbes de Provence

- Place the ingredients in a non-reactive container with a lid and leave to infuse, covered, for at least 24 hours.
- Once infused, strain through a muslin-lined sieve (strainer) or coffee filter and collect the liquid.
- Decant the liquid into a bottle and reserve in the fridge until required.

ROASTED COSMOPOLITAN MIX
Yields approximately 375 ml/13 fl oz

2 large pieces bone marrow
4 sprigs rosemary
250 ml/9 fl oz citron vodka
125 ml/4 fl oz triple sec
sea salt and freshly ground pepper, to season

- Preheat the oven to 200°C/400°/gas mark 6 and line a baking sheet with foil.
- Place the bone marrow on the baking tray, season with salt and pepper and roast in the oven for 20–25 minutes, until cooked through. Remove from the oven and set aside to rest.
- Preheat the water bath to 60°C/140°F.
- Place the bone marrow and the rest of the ingredients in a vacuum bag and seal.
- Place the vacuum bag in the water bath and leave to cook for 45 minutes, agitating the bag occasionally to infuse the flavours.
- Remove the vacuum bag from the water bath and set aside to cool.
- Once cooled, transfer the sealed bag to the freezer for 3–4 hours to allow the fat and solids to solidify and freeze.
- Carefully remove the bone marrow from the bag and discard, then strain the liquid through a muslin-lined sieve (strainer) or coffee filter and collect (this will take around 30 minutes).
- Decant the liquid into a bottle and reserve in the fridge until required.

BLACKCURRANT WRAY & NEPHEW
Yields approximately 125 ml/4 fl oz

125 ml/4 fl oz Wray and Nephew overproof rum
20 drops food-grade blackcurrant essence

- Place the rum and blackcurrant essence in an empty bottle or non-reactive container with a lid and seal.
- Roll the container to combine the liquids and reserve in the fridge until required.

COFFEE COCCHI
Yields approximately 350 ml/12 fl oz

10 g/½ oz spent coffee grinds
350 ml/12 fl oz Cocchi Americano

- Preheat the water bath to 45°C/113°F.
- Place the ingredients in a vacuum bag and seal.
- Place the vacuum bag in the water bath and leave to cook for 35 minutes.
- Remove the vacuum bag from the water bath and set aside to cool.
- Strain the infusion through a muslin-lined sieve (strainer) or coffee filter into a large jug or bowl.
- Decant the liquid into a bottle and reserve in the fridge until required.

COFFEE LIQUEUR
Yields approximately 400 ml/14 fl oz

100 g/3½ oz fresh-roasted coffee beans
500 ml/18 fl oz vodka
100 g/3½ oz caster (superfine) sugar

- Place the coffee beans and vodka in a non-reactive plastic container with a lid and leave to infuse, covered, for 48 hours.
- Once infused, strain the mixture through a muslin-lined sieve (strainer) or coffee filter and collect the liquid.
- Add the sugar and stir to combine. If you prefer a sweeter liqueur, add more sugar until you are happy with the taste.
- Decant the liqueur into a bottle and store in the fridge until required

COFFEE-INFUSED DARK RUM
Yields approximately 450 ml/16 fl oz

50 g/1¾ oz spent coffee grinds
500 ml/18 fl oz Bacardi Carta Negra Rum

- Place the coffee grinds and rum in a non-reactive plastic container with a lid and leave to infuse, covered, for 30 minutes.
- Once infused, strain the mixture through a muslin-lined sieve (strainer) or coffee filter and collect the liquid.
- Decant the liquid into a bottle and store in the fridge until required.

MILKYBAR LIQUEUR
Yields approximately 500 ml/18 fl oz

250 g/9 oz Milkybar (or other white chocolate) buttons
250 ml/9 fl oz vodka
45 g/1¾ oz milk powder
250 ml/9 fl oz water

- Fill a large pan one-third with water and place over a high heat. Once boiling, reduce to a gentle simmer.
- Place the chocolate buttons in a large heatproof mixing bowl and set over the pan, ensuring the base of the bowl is not touching the water. Stir continuously until the chocolate has completely melted, then remove the bowl from the pan and set aside.
- Pour the vodka, milk powder and 250 ml/9 fl oz of water into a jug and stir until well combined.
- Stirring continuously, pour the vodka mixture over the melted chocolate and continue to stir until well combined.
- Transfer the mixture to a non-reactive container with a lid and place in the fridge, covered, for at least an hour to allow the fat to separate from the chocolate liqueur.
- Once chilled, pass the mixture through a fine sieve (strainer) to remove the fatty solids and collect the liquid.
- Decant the liquid into a bottle and reserve in the fridge until required, for up to 5 days.

SEASONAL COGNAC
Yields approximately 350 ml/12 fl oz

350 ml/12 fl oz Cognac
300 g/10½ oz pear, finely sliced
125 g/4½ oz blackberries
2 cinnamon sticks, broken into shards

- Preheat the water bath to 45°C/113°F.
- Place the ingredients in a vacuum bag and seal.
- Place the vacuum bag in the water bath and leave to cook for 40 minutes.
- Remove the vacuum bag from the water bath and set aside to cool.
- Strain the infusion through a muslin-lined sieve (strainer) or coffee filter into a large jug or bowl.
- Decant the liquid into a bottle and reserve in the fridge until required.

BLACK OLIVE MARTINI BIANCO

Yields approximately 350 ml/12 fl oz

350 ml/12 fl oz Martini Bianco
3 g/0.1 oz food-grade black olive essence

- Place the Martini and olive essence in an empty bottle or non-reactive container with a lid and seal.
- Roll the container to combine the liquids and reserve in the fridge until required.

BAY, BIRCH & VANILLA VODKA

Yields approximately 350 ml/12 fl oz

350 ml/12 fl oz vodka
8 bay leaves
1 g/0.035 oz birch leaves
4 vanilla pods, halved lengthways

- Preheat the water bath to 45°C/113°F.
- Place the ingredients in a vacuum bag and seal.
- Place the vacuum bag in the water bath and leave to cook for 25 minutes.
- Remove the vacuum bag from the water bath and set aside to cool.
- Strain the infusion through a muslin-lined sieve (strainer) or coffee filter into a large jug or bowl.
- Decant the liquid into a bottle and reserve in the fridge until required.

CARAWAY TEQUILA

Yields approximately 250 ml/9 fl oz

250 ml/9 fl oz Cabeza Tequila (or another earthy variety)
7g/¼ oz caraway seeds

- Preheat the water bath to 45°C/113°F.
- Place the ingredients in a vacuum bag and seal.
- Place the vacuum bag in the water bath and leave to cook for 35 minutes.
- Remove the vacuum bag from the water bath and set aside to cool.
- Strain the infusion through a muslin-lined sieve (strainer) or coffee filter into a large jug or bowl.
- Decant the liquid into a bottle and reserve in the fridge until required.

BUTTER-WASHED TEQUILA

Yields approximately 600 ml/20 fl oz

700 ml/1¼ pints Patron Reposado tequila
500 g/1 lb 2 oz grass-fed unsalted butter, melted

- Preheat the water bath to 50°C/122°F.
- Place the ingredients in a vacuum bag and seal.
- Place the vacuum bag in the water bath and leave to cook for 55 minutes.
- Remove the vacuum bag from the water bath and set aside to cool. Once cooled, transfer the sealed bag to the freezer for 24 hours to allow the fats to freeze.
- Cut the corner of the vacuum bag and filter the infusion through a muslin-lined sieve (strainer) or coffee filter and collect the liquid (this will take around 30 minutes).
- Decant the liquid into a bottle and reserve in the fridge until required.

CARAMELIZED RED ONION REDUCTION
Yields approximately 250 ml/9 fl oz

200 g/7 oz caster (superfine) sugar
15 g/½ oz food-grade red onion extract
200 ml/7 fl oz vodka
1 g/0.035 oz sage leaves, torn

- Place the sugar in a pan over a medium heat and cook until syrupy and golden, do not stir the sugar as this will cause it to crystallize.
- Add the red onion extract to the pan and stir to combine, then remove the pan from the heat and carefully pour in half the vodka.
- Return the pan to a low heat and add the sage leaves. Continue to heat, stirring occasionally, until the caramel has melted into the vodka, then remove from the heat and set aside to cool.
- Once cooled, add the remaining vodka, then strain the liquid through a muslin-lined sieve (strainer) or coffee filter into a non-reactive container with a lid.
- Reserve in the fridge until required.

YUZU LIQUEUR
Yields approximately 500 ml/18 fl oz

50 g/1¾ oz yuzu peel
350 ml/12 fl oz vodka
1 g/0.035 oz citric acid powder
150 ml/5 fl oz sugar syrup (see page 26)

- Place the yuzu peel and vodka in a non-reactive container with a lid and leave to infuse, covered, for at least 6 hours.
- Once infused, strain through a muslin-lined sieve (strainer) or coffee filter and collect the liquid.
- Stir in the citric acid powder until dissolved and add the sugar syrup.
- Decant the liquid into a bottle and reserve in the fridge until required.

YUZU GIN
Yields approximately 350 ml/12 fl oz

50 g/1¾ oz yuzu peel
350 ml/12 fl oz gin

- Place the ingredients in a non-reactive container with a lid and leave to infuse, covered, for at least 6 hours.
- Once infused, strain through a muslin-lined sieve (strainer) or coffee filter and collect the liquid.
- Decant the liquid into a bottle and reserve in the fridge until required.

WINTER-SPICED LIQUEUR
Yields approximately 550 ml/18½ fl oz

600 g/1 lb 5 oz good-quality mincemeat
700 ml/1¼ pints gin

- Preheat the water bath to 45°C/113°F.
- Place the ingredients in a vacuum bag and seal.
- Place the vacuum bag in the water bath and leave to cook for 50 minutes, agitating the bag occasionally to infuse the flavours.
- Remove the vacuum bag from the water bath and set aside to cool.
- Once cooled, transfer the sealed bag to the freezer for 12 hours to allow the fats to freeze.
- Cut the corner of the vacuum bag and filter the infusion through a muslin-lined sieve (strainer) or coffee filter and collect the liquid (this will take around 30 minutes).
- Decant the liquid into a bottle and reserve in the fridge until required.

PINK PEPPERCORN GIN
Yields approximately 500 ml/18 fl oz

500 ml/18 fl oz gin
10 drops food-grade pink peppercorn essence

- Preheat the water bath to 45°C/113°F.
- Place the ingredients in a vacuum bag and seal.
- Place the vacuum bag in the water bath and leave to cook for 50 minutes.
- Remove the vacuum bag from the water bath and set aside to cool.
- Strain the infusion through a muslin-lined sieve (strainer) or coffee filter into a large jug or bowl.
- Decant the liquid into a bottle and reserve in the fridge until required. (The mixture will initially appear cloudy, but will clear within a few hours of decanting.)

ACORN LIQUEUR
Yields approximately 500 ml/18 fl oz

350 ml/12 fl oz Grey Goose vodka
5 g/⅛ oz food-grade acorn essence
150 g/5½ oz caster (superfine) sugar

- Preheat the water bath to 45°C/113°F.
- Place the vodka and acorn essence in a vacuum bag and seal.
- Place the vacuum bag in the water bath and leave to cook for 45 minutes.
- Remove the vacuum bag from the water bath and set aside to cool.
- Transfer the infusion to a non-reactive container with a lid and stir in the sugar until dissolved.
- Reserve in the fridge until required.

KOBE/WAGYU WHISKY
Yields approximately 650 ml/22 fl oz

100 g/3½ oz kobe or wagyu beef fat
700 ml/1¼ pints Japanese whisky

- Preheat the water bath to 45°C/113°F.
- Place a large frying pan or skillet over a medium heat. Once hot, add the beef fat and cook until completely rendered and melted.
- Place the whisky in a vacuum bag followed by the hot rendered fat (adding the fat first could damage the bag), then seal.
- Place the vacuum bag in the water bath and leave to cook for 50 minutes.
- Remove the vacuum bag from the water bath and set aside to cool.
- Once cooled, transfer the sealed bag to the freezer for at least 8 hours to allow the fats to freeze (the longer the freeze time, the clearer the end result).
- Cut the corner of the vacuum bag and filter the infusion through a muslin-lined sieve (strainer) or coffee filter and collect the liquid (this will take around 30 minutes).
- Decant the liquid into a bottle and reserve in the fridge until required.

BLUE CHEESE GIN
Yields approximately 700 ml/1¼ pints

300 g/10½ oz good-quality blue cheese, cut into 2 cm/¾ in cubes
700 ml/1¼ pints Bombay Sapphire gin

- Preheat the water bath to 52°C/126°F.
- Place the cheese and gin in a vacuum bag and seal.
- Place the vacuum bag in the water bath and leave to cook for 50 minutes.
- Remove the vacuum bag from the water bath and set aside to cool. Once cooled, transfer the sealed bag to the freezer for a minimum of 8 hours to allow the fats to freeze.
- Cut the corner of the vacuum bag and filter the infusion through a muslin-lined sieve (strainer) or coffee filter and collect the liquid (this will take around 30 minutes).
- Decant the liquid into a bottle and reserve in the fridge until required.

PINE NEEDLE GIN
Yields approximately 700 ml/ 1¼ pints

200 g/7 oz pine needles, washed
700ml/1¼ pints Hendrick's gin

- Place the ingredients in a non-reactive container with a lid and leave to infuse, covered, for at least 24 hours.
- Once infused, strain through a muslin-lined sieve (strainer) or coffee filter and collect the liquid.
- Decant the liquid into a bottle and reserve in the fridge until required.

DISTILLATIONS

RISTRETTO VODKA
Yields approximately 350 ml/12 fl oz

22 g/¾ oz fresh coffee beans
155 ml/5¼ fl oz fresh espresso
350 ml/12 fl oz vodka
mineral water, to rectify

- Preheat the water bath of your still to 50°C/122°F.
- Add the ingredients to your distilling flask and set over the water bath.
- Begin the distillation and gradually reduce the pressure to 50 Mbars.
- Once you have collected 350 ml/12 fl oz of liquid, stop the process.
- Rectify the distillation with mineral water to a strength of 40 % ABV (80 Proof).
- Decant the infused vodka into a bottle and reserve in the fridge until required.

WHOLEGRAIN MUSTARD VODKA
Yields approximately 500 ml/18 fl oz

75 g/2¾ oz wholegrain mustard
500 ml/18 fl oz vodka
mineral water, to rectify

- Preheat the water bath of your still to 55°C/131°F.
- Add the mustard and vodka to your distilling flask and set over the water bath.
- Begin the distillation and gradually reduce the pressure to 60 Mbars.
- Once you have collected 500 ml/18 fl oz of liquid, stop the process.
- Rectify the distillation with mineral water to a strength of 40 % ABV (80 Proof).
- Decant the infused vodka into a bottle and reserve in the fridge until required.

BRANSTON PICKLE VODKA
Yields approximately 500 ml/18 fl oz

215 g/7½ oz Branston Pickle
500 ml/18 fl oz vodka
mineral water, to rectify

- Preheat the water bath of your still to 60°C/140°F.
- Place the Branston Pickle and vodka in a blender and blitz to break down the larger chunks of pickle. Decant into your distilling flask and set over the water bath.
- Begin the distillation and gradually reduce the pressure to 50 Mbars.
- Once you have collected 500 ml/18 fl oz of liquid, stop the process.
- Rectify the distillation with mineral water to a strength of 40 % ABV (80 Proof).
- Decant the infused vodka into a bottle and reserve in the fridge until required.

PARSLEY VODKA
Yields approximately 400 ml/14 fl oz

350 ml/12 fl oz vodka
35 g/1¼ oz fresh flat-leaf parsley, washed
mineral water, to rectify

- Preheat the water bath of your still to 48°C/118°F.
- Place the vodka and parsley in a blender and blitz to combine. Decant into your distilling flask and set over the water bath.
- Begin the distillation and gradually reduce the pressure to 50 Mbars.
- Once you have collected 350 ml/12 fl oz of liquid, stop the process.
- Rectify the distillation with mineral water to a strength of 50 % ABV (100 Proof).
- Decant the infused vodka into a bottle and reserve in the fridge until required.

POLLEN VODKA
Yields approximately 700 ml/1¼ pints

700 ml/1¼ pints vodka
120 g/4½ oz pollen
mineral water, to rectify

- Preheat the water bath of your still to 55°C/131°F.
- Place the vodka and pollen in a blender and blitz to combine. Decant into your distilling flask and set over the water bath.
- Begin your distillation and gradually reduce the pressure to 45 Mbars.
- Once you have collected 500 ml/18 fl oz of liquid, stop the process.
- Rectify the distillation with mineral water to a strength of 40 % ABV (80 Proof).
- Decant the infused vodka into a bottle and reserve in the fridge until required.

PEANUT BUTTER LIQUEUR
Yields approximately 500 ml/18 fl oz

350ml/12 fl oz vodka
340 g/11¾ oz peanut butter
125 g/4½ oz caster (superfine) sugar

- Preheat the water bath of your still to 50°C/122°F.
- Place the vodka and peanut butter in a blender and blitz to combine. Decant into your distilling flask and set over the water bath.
- Begin the distillation and gradually reduce the pressure to 50 Mbars until you have collected 350 ml/12 fl oz of liquid. Then stop the process.
- Separately, place the sugar in a pan over a medium heat and cook until the sugar caramelizes. Do not stir the sugar as this will cause it to crystallize.
- Once caramelized, remove the pan from the heat and leave to cool slightly. Add in the peanut butter distillation and return to a low heat.
- Heat gently until the caramelized sugar has melted into the liquid, then remove the pan from the heat and leave to cool.
- Once cooled, filter the mixture through a muslin-lined sieve (strainer) or coffee filter and decant the liquid into a bottle.
- Reserve in the fridge until required.

SESAME SCOTCH DISTILLATE
Yields approximately 350 ml/12 fl oz

350 ml/12 fl oz blended Scotch whisky
15ml/½ fl oz toasted sesame oil
mineral water, to rectify

- Preheat the water bath of your still to 52°C/126°F.
- Place the ingredients in your distilling flask and set over the water bath.
- Begin the distillation and gradually reduce the pressure to 50 Mbars.
- Once you have collected 250 ml/9 fl oz of liquid, stop the process.
- Rectify the distillation with mineral water to a strength of 40 % ABV (80 Proof).
- Decant the infused whisky into a bottle and reserve in the fridge until required.

GREEN APPLE RYE
Yields approximately 700 ml/1¼ pints

700 ml/1¼ pints rye whiskey
250 g/9 oz green apples, roughly chopped
mineral water, to rectify

- Preheat the water bath of your still to 60°C/140°F.
- Place the whiskey and apples in a blender and blitz well to combine. Decant into your distilling flask and set over the water bath.
- Begin the distillation and gradually reduce the pressure to 50 Mbars.
- Once you have collected 500 ml/18 fl oz of liquid, stop the process.
- Rectify the distillation with mineral water to a strength of 50 % ABV (100 Proof).
- Decant the infused whiskey into a bottle and reserve in the fridge until required.

SHIITAKE VODKA
Yields approximately 350 ml/12 fl oz

35 g/1¼ oz dried shiitake mushrooms
350 ml/12 fl oz vodka
mineral water, to rectify

- Preheat the water bath of your still to 50°C/122°F.
- Separately, place the mushrooms and vodka in a blender and blitz to combine. Decant into your distilling flask and set over the water bath.
- Begin the distillation and gradually reduce the pressure to 60 Mbars.
- Once you have collected 250 ml/9 fl oz of liquid, stop the process.
- Rectify the distillation with mineral water to a strength of 40 % ABV (80 Proof).
- Decant the infused vodka into a bottle and reserve in the fridge until required.

HAY JACK DANIELS
Yields approximately 350 ml/12 fl oz

350 ml/12 fl oz Jack Daniels
8 g/¼ oz hay, washed
mineral water, to rectify

- Preheat the water bath of your still to 50°C/122°F.
- Place the whiskey and hay in a blender and blitz well to combine. Decant into your distilling flask and set over the water bath.
- Begin the distillation and gradually reduce the pressure to 50 Mbars.
- Once you have collected 250 ml/9 fl oz of liquid, stop the process.
- Rectify the distillation with mineral water to a strength of 40 % ABV (80 Proof).
- Decant the infused whiskey into a bottle and reserve in the fridge until required.

REMOVED KAHLUA
Yields approximately 300 ml/10 fl oz

700 ml/24 fl oz Kahlua
mineral water, to rectify

- Preheat the water bath of your still to 60°C/140°F.
- Decant the Kahlua into your distilling flask and set over the water bath.
- Begin the distillation and gradually reduce the pressure to 45 Mbars.
- Once you have collected 250 ml/9 fl oz of liquid, stop the process.
- Rectify the distillation with mineral water to a strength of 30 % ABV (60 Proof).
- Decant the infused vodka into a bottle and reserve in the fridge until required.

WASABI VODKA
Yields approximately 500 ml/18 fl oz

400 ml/14 fl oz vodka
45 g/1¾ oz wasabi
mineral water, to rectify

- Preheat the water bath of your still to 55°C/131°F.
- Place the vodka and wasabi in a blender and blitz well to combine. Decant into your distilling flask and set over the water bath.
- Begin the distillation and gradually reduce the pressure to 55 Mbars.
- Once you have collected 350 ml/12 fl oz of liquid, stop the process.
- Rectify the distillation with mineral water to a strength of 40 % ABV (80 Proof).
- Decant the infused vodka into a bottle and reserve in the fridge until required.

SUPPLIERS

BAR AND SPECIALIST EQUIPMENT

Cocktail Kingdom
A great collection of bar equipment for delivery worldwide – from shakers and strainers to ice moulds and bar spoons. Checkout their glassware range, too. They also have a New York-based showroom.
www.cocktailkingdom.com
+1 (212) 647 9166

Nisbets
Previously a go-to supply shop for all manner of bar and kitchen supplies – now an extensive online shop with international shipping. From sous-vide and vacuum machines to larger kitchen and bar equipment. They also supply a great range of glassware.
www.nisbets.com
+44 (0)845 140 5555

Sous Vide Tools
From smokers, vacuum sealers, thermal circulators and everything in between, these guys have it all. UK-based but ship worldwide.
www.sousvidetools.com
+44 (0)800 678 5001

Cream Supplies
For molecular service ware and all your barista needs. This online shop is a heaven to browse and also stocks flavour enhancers, emulsifiers and other powders. They also have the odd rotary evaporator online.
www.creamsupplies.co.uk
+44 (0)23 923 78700

Molecular Recipes
US company stocking equipment from dehydrators and moulds to whippers and service ware. Ships worldwide.
www.molecularrecipes.com

Fancy Straws
The name says it all. Any colour and most patterns.
www.fancystraws.co.uk
+44 (0)1485 579363

GLASSWARE

Artis
www.artis-uk.com
+44 (0)20 8391 5544

Schott Zwiesel
www.shopstyle.co.uk

Riedel
www.riedel.co.uk
+44 (0)1782 646105

FOOD-GRADE ESSENCES AND DRY GOODS

Aftelier
For essences and perfumes. This US-based company ship and have some great products.
www.aftelier.com
+1 (510) 841 2111

Just Ingredients
Most of the dried herbs and spices that are used in the recipes within this book can be bought here. UK-based but ship worldwide.
www.justingredients.co.uk

Healthy Supplies
Dried seed and nuts, spices and herbs. UK-based.
www.healthysupplies.co.uk
+44 (0)800 689 1982

G. Baldwin & Co.
A south London-based company specializing in natural dried goods including flowers, hops and botanicals. I also pick up my pipette and bitter bottles here.
www.baldwins.co.uk
+44 (0)20 7703 5550

Foodie Flavours
Stocks a large range of natural food-grade flavourings, from blueberry to cream soda and dandelion to ginger. UK-based.
www.foodieflavours.com
+44 (0)333 222 5968

INDEX

WITH THANKS ...

Over the years, I have been influenced by many and by much – both directly and indirectly. As always, there are so many people who have helped to steer the direction in which I travel. If I have missed anyone, it is certainly not intentional and to those, I apologise.

In my recent years whilst at Duck & Waffle and SUSHISAMBA, I have been unbelievably fortunate to work with some amazing and talented individuals, who make up two separate but equally dedicated teams.

To Dan Doherty, Tom Cenci, Dan Barbosa, James Kirk-Gould, Ben Leek and all the chefs (both past and present) at Duck & Waffle who have been a joy to work alongside.

To Claudio Cardoso, Andreas Bollanos and the chefs at SUSHISAMBA.

To Daniel Susko, Bernardas, Massimiliano, Pietro, Francesco Virag, Steffen and the bar team at Duck & Waffle and the bar teams at the newly opened Duck & Waffle Local as well as the SUSHISAMBA locations in London, New York, Miami and Las Vegas – for continuing to allow me to curate ever changing menus and for being patient in my endeavours to push further.

To Gyunay AKA the prep guy. Who has the enduring pleasure of working with me on an almost daily basis and who continuously ensures high standards. He may spend most of his time behind the scenes, but my job would be a hell of a lot harder without him and his efforts.

To Shimon Bokovza, the visionary behind SUSHISAMBA and Duck & Waffle restaurants and a man who has more creative ideas in one day than most of us will ever have in a lifetime. Thank you for allowing me to join your company and for having the time, patience and understanding, to show you what is possible.

For Brian Bendix. You were undoubtedly my first personal industry mentor and someone whom I greatly respect. You allowed me to take that all important first step from operational roles (which is what I should have been doing), to be as creative as I allowed myself to be.

To friends – industry and otherwise. And to those who helped to inspire my direction (both directly and indirectly), and to those who have given advice and to those who continue to do so. Most notably…
Becci & Ade Broome, Marc Plumridge, Tom Hobbs, Mat Sayer, Jacob Briars, Ben Reed, Tony Conigliaro, Ryan Chetiyawardana, Matt Whiley, Clerkenwell Boy, Sandrae and Gary (The Cocktail Lovers), May Face and Shev.

For those who made this book possible:
Shy Lewis (Shy PR), You've always been the most patient of agents, but have become a great adviser, friend and sounding board. Here's to the next chapter …

To Katie, Fiona, Dan & Laura from Pavilion Books. Thanks for allowing me the creativity and freedom to see my ideas through. To Liz and Max Haarala-Hamilton for the most beautiful of imagery and capturing my ideas and to Jack Sargeson for the great styling of them. Also to Valerie Berry and Alex Breeze for their food and prop styling, and Lucy and Mike who provided the great shoot location.

And, finally…

For my mum – Julie. A brave and strong woman who I have so much love and respect for. You taught me the ethics I have today and gave me so many opportunities growing up – I hope I make you proud.

For my daughter, Isabella-Rose. By far and away my proudest creation.

And, lastly, for Yvette – the inspiration behind many of my icon drinks, and the true creator of the Nutella Negroni. No drink ever tastes as good as those shared with you.

'If I have seen further than others, it is by standing upon the shoulders of giants…'
Isaac Newton

First published in the United Kingdom in 2017
by
Pavilion
43 Great Ormond Street
London
WC1N 3HZ

ISBN 978-1-91121-614-8

A CIP catalogue record for this book is available
from the British Library.

10 9 8 7 6 5 4 3 2 1

Reproduction by Rival Colour Ltd, UK
Printed by Toppan Leefung Printing Ltd, China

This book can be ordered direct from the
publisher at www.pavilionbooks.com